The Regulation of Cyberspace

Control in the Online Environment

In *The Regulation of Cyberspace* Andrew Murray examines the development and design of regulatory structures in the online environment.

The book considers and models how all forms of control – including design and market controls, as well as traditional command and control regulation – are applied within the complex and flexible environment of cyberspace. Drawing on the work of cyber-regulatory theorists such as Yochai Benkler, Andrew Shapiro and Lawrence Lessig, *The Regulation of Cyberspace* suggests a model for cyberspace regulation which acknowledges its complexity. It further suggests how this model can be utilised by regulators to provide a more comprehensive regulatory structure for cyberspace.

Andrew D. Murray is Senior Lecturer in IT and Internet Law in the Department of Law, London School of Economics.

The Regulation of Cyberspace

Control in the Online Environment

Andrew D. Murray

Routledge·Cavendish
Taylor & Francis Group
a GlassHouse book

First published 2007 by Routledge-Cavendish
2 Park Square, Milton Park, Abingdon, Oxon OX14 4RN

Simultaneously published in the USA and Canada
by Routledge-Cavendish
270 Madison Ave, New York, NY 10016

A Glasshouse book

*Routledge-Cavendish is an imprint of the Taylor & Francis Group, an
informa business*

© 2007 Andrew D. Murray

Typeset in Times New Roman by RefineCatch Limited, Bungay,
Suffolk
Printed and bound in Great Britain by
TJ International, Padstow, Cornwall

British Library Cataloguing in Publication Data
A catalogue record for this book is available from the British Library

Library of Congress Cataloging in Publication Data
Murray, Andrew D.
   The regulation of cyberspace : control in the online environment /
Andrew D. Murray. — 1st ed.
     p. cm.
   ISBN 1–904385–21–4 (hbk.)
1. Internet—Law and legislation.   1. Title.
   K564.C6M87 2006
   343.09′944—dc22   2006020898

ISBN10: 0–415–42001–6 (pbk)
ISBN10: 1–904385–21–4 (hbk)

ISBN13: 978–0–415–42001–3 (pbk)
ISBN13: 978–1–904385–21–9 (hbk)

For Mum & Dad
They are the most influential people in my life and they made me the person I am today.
And Rachel
For her patience.

Contents

Preface

One of the most impassioned and vibrant academic debates in the last decade has been the intricate question of whether it is possible to regulate the actions of individuals in cyberspace and if so, what form such regulation should take. I have followed this debate throughout its development and have on occasion contributed to it. Despite this I have always thought of myself as somewhat of an outsider in this debate as, to date, full membership of the community of cyber-regulatory theorists and with it the right to fully participate seems to be an almost exclusive preserve of the North American *cyberlawyer*: a small group of academics who possess a unique blend of legal-regulatory training and technical know-how. A literature review of the field demonstrates that it is dominated by the works of US scholars such as David Post, David Johnson, Joel Reidenberg, Milton Mueller, Yochai Benkler, Pamela Samuelson, Michael Froomkin, Mark Lemley and, of course, Lawrence Lessig. Should you seek out a European view on cyber-regulation you find that European scholars are divided into two categories, neither of which reflect the US approach. Some, such as Jon Bing, Chris Reed, Richard De Mulder, Colin Tapper and Peter Blume follow a normative approach in their analysis of *computer law*. They choose to study the disruptive effects of computers and computerised processes on the established legal order; they do not, though, seek to extend this analysis into regulatory theory or systems of control. The second category consists of those scholars such as Manuel Castells, Robin Mansell, Annabelle Sreberny, Sonia Livingstone and Edward Steinmueller, who conduct research into the social effects of digital technology from a communications and media standpoint, but who take a limited interest in regulation and regulatory theory. As quickly becomes clear there is no European (or for that matter Antipodean, Asian or African) counterpoint to the North American analysis: the whole debate is perilously parochial. In an attempt to develop a distinctively European voice in this debate I took my reading into a slightly different direction. I began to read European and Commonwealth commentators on regulatory theory in the hope that by immersing myself in their culture, I would discover the points of contact between regulatory theory in general and cyber-regulatory theory in particular. To this end I read works by John Braithwaite and Peter Drahos on business regulation, Robert Baldwin and Martin Cave on regulatory systems and utilities regulation, and Clifford Shearing and Julia Black on decentred or nodal regulation. They demonstrated that to have a thorough grasp of regulatory theory requires an understanding of complex relationships including, but not limited to, business relationships, economic or market relationships, relationships of power, and social or community interaction. In so doing they demonstrated that from a regulatory theory standpoint there is nothing particularly unusual about cyber-regulation. Rather, it may be argued that cyber-regulation and cyber-regulatory theory shows a disproportionate regard for the role of technology in regulation and regulatory relationships, or to put it another way, cyber-regulatory theorists suffer from excessive techno-determinism.

With this as my starting point I began to explore some basic questions in cyber-regulatory theory from my unique vantage point. This book tells the story of this journey and hopefully in so doing develops a uniquely European contribution to the study of cyber-regulation and cyber-regulatory theory. It is split into three sections, each of which reflects a key part of the journey and its final destination. The first part – *Mapping the cyber-regulatory environment* – reflects an initial period of uncertainty and darkness in the journey. After reading the work of Braithwaite, Drahos, Cave and Black, I asked whether there was value in studying cyber-regulation, as distinct from regulation. I found I was not the first to question the value of cyber-regulatory theory and in this section I examine the arguments for and against the study of cyber-regulatory theory as distinct from regulatory theory. In answering this question I found that I came up against a new problem. My research into regulation and complexity threw up a further issue of whether it is possible to map cyber-regulatory patterns. Eminent scientists and statisticians have posited the problem of *chaos* for over half a century and the work of leading chaos theorists including Ludwig von Bertalanffy and Edward Lorenz suggested to me that attempts to map regulatory patterns in complex systems such as cyberspace would ultimately prove fruitless. This suggested that any cyber-regulatory analysis would be undermined by the uncertainty chaos brings. Fortunately, the contemporary field of synchronicity offered tantalising glimpses at a solution. A chance visit to the Science Museum in London offered some treasures including John Gribbin's *Deep Simplicity* and Steven Strogatz's *Sync: the emerging science of spontaneous order*. Spurred on by their analyses, which showed that even the most chaotic systems demonstrate patterns if viewed over time, I decided to press on. In the second part – *Regulatory tools and digital content* – over five chapters loosely based on Lawrence Lessig's four modalities of regulation (architecture, norms, markets and laws) I analyse what causes regulatory interventions to succeed or to fail in cyberspace in anticipation of uncovering valuable patterns, which may be harnessed by regulators to design effective regulatory tools within the digital environment (or indeed any complex environment). The result of this analysis is outlined in the final part – *Regulating cyberspace: Challenges and opportunities*. Here I suggest that by adopting dynamic regulatory models, regulators may design *symbiotic regulation*: regulation which by harnessing the pre-existing lines of communication within the extant regulatory environment are sympathetic to that environment, giving it a far greater likelihood of success. These models suggest that with some simple models to map the dynamics of communication within a regulatory environment, regulators may regulate effectively within that environment without the need to map every potential system variation caused by their intervention. In effect it is an application of the principles of synchronicity, which suggest that harmony arises naturally and may be harnessed, in the socio-regulatory environment. With this conclusion I completed my journey. I hope by following this journey the reader, like myself, encounters ideas and concepts that challenge traditional assumptions and provides opportunities to develop a more thorough understanding of the regulatory environment both within cyberspace and of more general application in the study of regulatory theory.

All that remains is for me to thank all those who provided guidance or support throughout my journey. To Julia Black for introducing me to nodal governance, Francis Snyder and Robert Reiner who asked difficult questions, Colin Scott who invited me to challenge the orthodox views of cyber-regulatory theory and Mathias Klang who tirelessly discussed my ideas and read my drafts and who saved me from many a faux pas. Particular thanks to all the students of my New Media classes who helped me focus these ideas through hours of class discussion. In particular (but in no order), Brian Esler, Luke Gibbs, Anthony Hamelle, Marina Amoroso, Anne Reisinger, Russ Taylor, Darcie Sherman, Max Mittasch and Carlisle George. Above all, thanks to those whose support made this possible. To my wife Rachel who has lived with my journey for the last two years, who has provided love and encouragement when most needed and who has assiduously read drafts of the text even as I redrafted them. Finally, thanks to my parents Andrew and Sarah Murray who invested their time and money to support me in undertaking my undergraduate studies in law at the University of Edinburgh where this journey really began. I hope this in some small way repays that investment.

Andrew D. Murray
31 May 2006

Acronyms and abbreviations

Chapter 1

C&IT	computers and information technology
EFF	Electronic Frontier Foundation
HTML	hypertext markup language
HTTP	hypertext transfer protocol
IETF	Internet Engineering Task Force
IP	internet protocol
IS	information systems
ISP	internet service provider
LAN	local area network
MUD	multi-user dungeon
SMTP	simple mail transfer protocol
TCP	transmission control protocol
WWW	World Wide Web

Chapter 2

CPC	Communist Party of China
DMCA	Digital Millennium Copyright Act
FRC	Federal Radio Commission
OSI	open systems interconnection
PRC	People's Republic of China
STL	socio-technical-legal

Chapter 3

ACM	Association for Computer Machinery
ARPA	Advanced Research Projects Agency
ARPANET	Advanced Research Projects Agency Network
B2C	business to consumer
GPS	Global Positioning System

GUI	graphical user interface
IMP	interface message processor
INWG	International Network Working Group
IP	internet protocol
IPTO	Information Processing Techniques Office
IRC	internet relay chat
ISO	International Organisation for Standardisation
JANET	Joint Academic NETwork
MUD	multi-user dungeon
NAP	network access points
NCP	network control protocol
NPL	National Physical Laboratory
OSI	open-system interconnection
RFID	radio frequency identification
STL	socio-technical-legal
TCP	transmission control protocol
WAP	wireless application protocol
WWW	World Wide Web

Chapter 4

ADR	alternate dispute resolution
ADSL	asymmetric digital subscriber lines
ccTLD	country code top-level domain
CDS	Cactus Data Shield
DCA	Defense Communications Agency (US)
DISA	Defense Information Systems Agency (US)
DNS	Domain Name System
DNSO	Domain Names Supporting Organisation
DOC	Department of Commerce (US)
DoD	Department of Defense (US)
eDNS	Enhanced Domain Name System
EU	European Union
FCC	Federal Communications Commission
FTP	file-transfer protocol
GAC	Government Advisory Committee
gTLD	generic top-level domain
HTML	hypertext markup language
HTTP	hypertext transfer protocol
IAB	Internet Architecture Board
IAHC	International Ad-Hoc Committee
IANA	Internet Assigned Numbers Authority
ICANN	Internet Corporation for Assigned Names and Numbers
ICT	information and communications technology

IDNs	Internationalised Domain Names
IESG	Internet Engineering Steering Group
IETF	Internet Engineering Task Force
IGF	Internet Governance Forum
IMP	interface message processor
IP	internet protocol
IPR	intellectual property rights
IRTF	Internet Research Task Force
ISP	internet service provider
ITU	International Telecommunications Union
MMS	Microsoft Media Server
MOU	memorandum of understanding
NDSS	Network and Distributed System Security
NGO	non-government organisation
NSI	Network Solutions Inc.
OS	operating system
OSI	open-system interconnection
PDA	personal digital assistant
PDF	Portable Document Format
R&D	research and development
RFC	request for comments
RIR	Regional Internet Registry
SMTP	simple mail transfer protocol
SRI	Stanford Research Institute
TCP	transmission control protocol
TLD	top-level domain
TPM	technical protection measures
UDRP	uniform domain-name dispute-resolution policy
UMTS	universal mobile telecommunications system
UN	United Nations
VoIP	voice over internet protocol
WGIG	Working Group on Internet Governance
WIPO	World Intellectual Property Organisation
WSIS	World Summit on the Information Society
WWW	World Wide Web

Chapter 5

BBS	bulletin board systems
CC	Creative Commons
FTP	file-transfer protocol
MUD	multi-user dungeon
P2P	peer-to-peer
RFC	request for comments

Chapter 6

ARPANET	Advanced Research Projects Agency Network
C2C	consumer-to-consumer
CC	Creative Commons
CSS	Content Scrambling System
DMCA	Digital Millennium Copyright Act
DRM	Digital Rights Management
EFF	Electronic Frontier Foundation
EU	European Union
ICT	information and communications technology
IFPI	International Federation of the Phonographic Industry
LSE	London School of Economics
MPAA	Motion Picture Association of America
OS	operating system
P2P	peer-to-peer
PKE	public-key encryption
RIAA	Recording Industry Association of America
SSL	Secure Socket Layer technology
VCR	video cassette recorder
WCT	WIPO Copyright Treaty

Chapter 7

CDA	Communications Decency Act (US)
DPP	Director of Public Prosecutions
EFF	Electronic Frontier Foundation
EU	European Union
IETF	Internet Engineering Task Force
NCROPA	National Campaign for the Reform of the Obscene Publications Acts
NVALA	National Viewers and Listeners Association
NYSPC	New York State Penal Code (US)
OPA	Obscene Publications Act

Chapter 8

CDA	Communications Decency Act (US)
DMCA	Digital Millennium Copyright Act
DOC	Department of Commerce (US)
DRM	Digital Rights Management
DVD	digital versatile disc
EU	European Union
ICANN	Internet Corporation for Assigned Names and Numbers

IGF	Internet Governance Forum
IPR	intellectual property rights
P2P	peer-to-peer
PRC	People's Republic of China
UN	United Nations
VCR	video cassette recorder
WIPO	World Intellectual Property Organisation
WSIS	World Summit on the Information Society

Table of cases

Table of statutes

Part I

Mapping the cyber-regulatory environment

Chapter 1

Coffee pots and protocols: The role of the cyberlawyer

Cyberlaw and the law of the Internet are not useful concepts.

Joseph Sommer

In the spring of 1994 I was a final year undergraduate student at the University of Edinburgh. One afternoon, while working in the library, Duncan, a friend and fellow student who was also a part-time software developer, suggested we visit the student microlab. The microlab was the Law School's collection of IBM PCs and Apple Macs, all networked to form a local area network or LAN, and used mostly by students for word processing and instant messaging with friends. He said he had something new and exciting to show me: a wonderful new device that had just been installed. Despite Duncan's obvious enthusiasm, I was quite sceptical. Duncan was an excellent software engineer and tended to get terribly excited by the most mundane of items such as a new PC chipset or an upgrade to the operating system. I saw myself as a lawyer not a computer scientist; as such I was rarely enthralled by such technology. Still, that afternoon he convinced me to come along. Once we were in the lab he logged on and brought up a grainy black and white picture on the monitor. I struggled to make sense of what I was seeing but after a few seconds I discerned that I was looking at a half-full filter coffee pot. I asked what was so exciting about this image. He told me I was looking at a live picture of a coffee pot in the computer laboratory at the University of Cambridge.[1] At first this didn't really make sense to me. How could I be

---

1 This is the now famous Trojan Room Coffee Pot. The coffee pot was first broadcast across a closed network at the computer laboratory in 1991. A small programme called XCoffee was developed by Paul Jardetzky, allowing researchers throughout the building to see if there was coffee in the pot before leaving their office for a refill. On 22 November 1993 this closed network went public when Daniel Gordon and Martyn Johnson updated the frame grabber and put the coffee pot on the fledgling World Wide Web – making it the first ever webcam. The coffee pot remained online until 22 August 2001, after which the £25 pot was sold for £3,350 via an internet auction site. A non-technical biography of the coffee pot by Quentin Stafford-Fraser is available at: *www.cl.cam.ac.uk/coffee/qsf/coffee.html*.

looking at a coffee pot in Cambridge? Duncan then explained to me that the computers in the microlab had recently been installed with a new piece of software called Mosaic, which allowed access to a new area on the internet called the World Wide Web. He went on to explain the basics of the web and as soon as his explanation was over I was hooked. This simple action of showing me that I could look at a coffee pot 400 miles away by typing a few characters changed the way I thought about computers and technology. I, like many millions before and since, became an instant devotee of the web.

Like most users of the web I was won over by its simplicity. Here was a truly accessible internet protocol (IP) for the non-technically minded. This simplicity led to an explosion of internet connectivity in North America and Western Europe throughout the 1990s.[2] I became a regular user of the web: it provided an excellent research tool for a young academic embarking on a fledgling career; it allowed me to keep in touch easily with my friends after graduation; and it provided an easily accessible source of entertainment. I didn't consider it to be anything more than a useful tool. I certainly never considered at the outset of my academic career in 1996 that there was anything of particular interest to lawyers and regulators. I, like many others at that time, saw the web as a particularly sophisticated telecommunications tool and little else. It allowed users to enter into one-to-many communications through personal web pages, many-to-many communications through chatrooms and MUDs,[3] and through email it allowed simple one-to-one communications.[4] Thus the internet, and the web in particular, allowed us to carry out in a more sophisticated manner, the types of social transactions we had been carrying out for years through broadcasting, everyday social interaction and narrowcast telecommunications systems such as the telephone and the mail system. I thought, as most lawyers at that time were thinking, that the role of the regulator in such technology was limited to enabling the technology to function and to control antisocial practices. I saw regulators as

2   Of course the development of the web, based upon Tim Berners-Lee's hypertext transfer protocol (HTTP) is only part of this tale. The strong growth in internet usage in the 1990s is also due in part to the opening of the internet to commercial traffic by the US National Science Foundation in 1991 and the adoption globally of the TCP/IP and SMTP protocols in the late 1980s and early 1990s. See Hafner, K and Lyon, M, *Where Wizards Stay Up Late: The Origins of the Internet*, 1996, New York: Touchstone, ch 8.
3   A MUD or multi-user dungeon is a space where users can engage in discussion or interactive gaming with other members of the dungeon. MUDs grew around the science fiction and dungeons and dragons communities of the 1980s. They provided text-based interactive entertainment. For more details on MUDs see Turkle, S, *Life on the Screen: Identity in the Age of the Internet*, 1995, New York: Simon & Schuster.
4   Most email systems use the simple mail transfer protocol (SMTP) and are therefore not part of the web (which is based on HTTP and HTML). Some webmail systems such as Hotmail interface the two and are therefore part of the web. This is discussed in greater detail in Hafner & Lyon, fn 2 at ch 7.

having limited roles such as licensing ISPs, approving technical standards, consumer protection and content control.[5]

The early adaptors

Fortunately, not all academic lawyers were thinking this way. In the United States a body of research had grown up around the nascent subject of *cyberlaw*. Some early cyberlaw theorists saw the potential of the internet, and the web, to provide an independent and unregulated social sphere. To these cyber-libertarian theorists, cyberspace was akin to the old West. It was a place where individual freedom was secured by the environment of the space, where government interference was, by design, minimal, and where standards, norms and later laws (if any) would derive from the collective will of the citizens of cyberspace. The unlikely totem of the cyber-libertarian school was an ex-lyricist for the Grateful Dead named John Perry Barlow. As a young man Barlow took over a cattle ranching operation called Bar Cross Land and Livestock in Cora, Wyoming and in 1971 at the age of 24 he became a lyricist for the Grateful Dead. Barlow was both politically astute and a first-rate businessman, and when he sold his ranch in 1988 he threw his considerable energy into something that had become increasingly important to him: examining and questioning the role of technology and networks in society. He announced his arrival on the network scene in 1990 with two historical interventions. The first of these made Barlow famous within the technical community, but the second was more important to the question of the structure and values of the network. First, Barlow sealed his position in the internet story by becoming the first person to use William Gibson's science fiction term cyberspace to describe the existing global electronic social space.[6] More importantly though, in July 1990, he, along with the creator of Lotus 1–2–3 Mitch Kapor, formed the Electronic Frontier Foundation (EFF). The EFF

5   The need for content control was immediately apparent. One of the most adaptable industries is the adult entertainment industry. It was a first mover/adaptor in the introduction of printing, photography, cinematography, video, home computing, bulletin board systems, the web and most recently third-generation mobile phone technology. The relationship between technology and the sex industry is discussed in Van Der Leun, ' "This is a Naked Lady": Behind Every New Technology is . . . Sex?', *Wired 1.01* March/April 1993. As a result, with every new technology there has been an established reaction that content control quickly follows. For a discussion of censorship in the movie industry see Miller, F, *Censored Hollywood: Sex, Sin, and Violence on Screen*, 1994, Paducah KT: Turner. See also the effective, Video Recordings Act 1984 (videotapes) and the abortive Communications Decency Act (internet communications).

6   Cyberspace is a term first used by William Gibson in his 1984 novel *Neuromancer*, London, Voyager/Harper-Collins,1995. In this tale characters moved around inside a computer generated landscape that was stable, populated, easily navigated, and the size of a country, maybe larger. He called this realm cyberspace.

was central to the development of the cyber-libertarian ethos. The founders of the EFF believed that governments and corporations would seek to control how this new technology would be used by individuals. The aim of the EFF was to seek to protect individual freedom against such intervention in the developing cyberspace. The role of the EFF was therefore to 'defend our rights to think, speak, and share our ideas, thoughts, and needs using new technologies, such as the internet and the World Wide Web'.[7] The EFF quickly set to work. Its first action was to assist a small games book publisher from Austin, Texas, named Steve Jackson Games, and several of the company's bulletin board users in raising an action against the United States Secret Service claiming unlawful search and seizure.[8] It followed this by assisting University of California PhD student Daniel Bernstein raise an action against the State Department claiming the restriction on the publication of encryption codes under the United States Munitions List was an unconstitutional restriction of his right to free expression.[9] In both cases the EFF was successful, establishing important principles and freedoms for the internet community as a whole. Their work started to attract the attention of the legal community, and in particular two law professors, David Johnson and David Post. Johnson and Post embraced the cyber-libertarian ethos, which linked enthusiasm for electronically mediated forms of living with libertarian ideas on freedom, society and markets. They, like Mitch Kapor and John Perry Barlow, believed that many fundamental freedoms were inherently protected in cyberspace:[10] that the inherent design features of the internet would render any attempts at state intervention futile.[11]

The high point for the cyber-libertarian thesis was in early 1996. That spring two key cyber-libertarian papers were published. On 8 February, Barlow published his now infamous *A Declaration of Independence of Cyberspace*. This

---

7 Taken from 'About the EFF' Available at: *www.eff.com/about*.

8 This case established the key principle that electronic mail deserves at least as much protection as a telephone call. See *Steve Jackson Games Inc v US Secret Service* 816 F Supp 432 (WD Tex. 1993) *affd Steve Jackson Games Inc v US Secret Service* 36 F 3d 457 (5th Cir. 1994).

9 *Daniel J Bernstein v United States Department of State et al.* No. C–95–0582 MHP (ND Ca., 15 April 1996) *affd Bernstein v USDOJ* 176 F 3d 1132 (9th Cir. 1999). It should be noted that litigation between the now Professor Bernstein and the US Government remains ongoing.

10 This school may more correctly be referred to as classical cyber-libertarianism. Following the publication of Lawrence Lessig's *Code and Other Laws of Cyberspace*, 1999, New York: Basic Books, the cyber-libertarian school recognised the inherent regulability of cyberspace and developed a neo-cyber-libertarian school that calls for the application of libertarian values within cyberspace and for the absence of state intervention within this sphere. See Post, D, 'What Larry Doesn't Get: Code, Law and Liberty in Cyberspace' 2000, 52 Stan LR 1439.

11 This is most famously put in John Perry Barlow's *A Declaration of Independence for Cyberspace* where he said to the 'Governments of the Industrial World . . . You have no sovereignty where we gather . . . You have no moral right to rule us *nor do you possess any methods of enforcement* we have true reason to fear'. Available at: *www.eff.org/~barlow/ Declaration-Final.html*.

coalesced the cyber-libertarian belief in the unregulability of bits:[12] with Barlow claiming that real-world governments would find it impossible to regulate within 'sovereign' cyberspace.[13] Then in May the Stanford Law Review published Johnson and Post's key paper *Law and Borders – The Rise of Law in Cyberspace*.[14] Here, they laid out for the first time a legal interpretation of the classical cyber-libertarian contention that regulation founded upon traditional state sovereignty, based as it is upon notions of physical borders, cannot function effectively in cyberspace. Instead, they argued, individuals may move seamlessly between zones governed by differing regulatory regimes in accordance with their personal preferences. Simply put, they viewed the internet as a medium that would foster regulatory arbitrage and undermine traditional hierarchically structured systems of control. Noting that 'control' emanates at the level of individual networks, they proposed that although forms of hierarchical control might be exerted over specific networks, the aggregate range of such rule sets was unlikely to lead to any form of centralised control of cyberspace. Accordingly, the 'Law of Cyberspace' would largely be determined by a free market in regulation in which network users would be able to choose those rule sets they found most congenial. Johnson and Post maintained that the various dimensions of internetworking could be governed by 'decentralised, emergent law' wherein customary and privately produced laws, or rules, would be produced by decentralised collective action leading to the emergence of common standards for mutual co-ordination.[15] In other words, they believed that the decentralised and incorporeal nature of cyberspace meant that the only possible regulatory system was one that developed organically with the consent of the majority of the citizens of cyberspace.[16] Their views were, though, about to be challenged by the emergence of a strong counter-argument within the legal community: the development of the cyber-paternalist.

12 Bits are representative of digital switches. They are best understood as an instruction to a computer to do or not do a particular action. The best description of a bit is given by Nicholas Negroponte in his book *Being Digital*, 1995, Chatham: Hodder & Stoughton, where he describes it as having 'no colour, size or weight . . . it is the smallest atomic element in the DNA of information . . . for practical purposes we consider a bit to be a 1 or a 0.' at p 14.

13 Barlow claims that 'Your legal concepts of property, expression, identity, movement, and context do not apply to us. They are all based on matter, and there is no matter here. See above, fn 11.

14 48 Stan LR 1367 (1996).

15 This notion parallels the concept of polycentric or non-statist law. See Bell, T, 'Polycentric Law' 1991/92, 7(1) *Humane Studies Review* 4; Bell, T, 'Polycentric Law in the New Millennium.' 1998, paper presented at The Mont Pelerin Society: 1998 Golden Anniversary Meeting, at Alexandria, Virginia. Available at: *www.tomwbell.com/writings/FAH.html*.

16 Johnson, D and Post, D, above fn 14. See also Johnson, D and Post, D, 'The New 'Civic Virtue' of the Internet: A Complex Systems Model for the Governance of Cyberspace' 1998, in *The Emerging Internet* (1998 Annual Review of the Institute for Information Studies) Firestone, C (ed).

That same year cyber-libertarian views were seriously challenged for the first time by the publication of a series of papers. Several commentators noted that its proponents appeared to base their arguments on an oversimplified understanding of social and political phenomena, and that they adopted a particularly right-wing view of regulatory systems. Early critics included Langdon Winner of the Rensselaer Polytechnic Institute,[17] Reilly Jones of the Extropy Institute,[18] and Joel Reidenberg of Fordham Law School. Despite sympathising with the view that internetworking leads to the disintegration of territorial and substantive borders as key paradigms for regulatory governance, Reidenberg argued that new models and sources of rules were being created in their place. To this end, he identified two distinct regulatory borders arising from complex rule-making processes involving States, the private sector, technical interests, and citizen forces. Each of these borders was seen as establishing the defining behavioural rules within their respective realms of the networking infrastructure. The first set of borders encompassed the contractual agreements among various internet service providers (ISPs). The second type of border was the network architecture. The key factor at this level, he claimed, were the technical standards because they establish default boundary rules that impose order in network environments. Reidenberg's key contribution to the evolving debate at this stage though was his conceptualisation of a 'Lex Informatica'.[19] This draws upon the principle of Lex Mercatoria and refers to the 'laws' imposed on network users by technological capabilities and system design choices. Reidenberg asserted that, whereas political governance processes usually establish the substantive laws of nation states, in Lex Informatica the primary sources of default rule-making are the technology developer(s) and the social processes through which customary uses of the technology evolve.[20] To this end, he argued that, rather than being inherently unregulable due to its design or architecture, the internet is in fact regulated by its architecture. Therefore, Lex Informatica could be seen as an important system of rules analogous to a legal regime. According to this view, internet-related conflicts and controversies reflect a state of flux in which Lex Informatica and established legal regimes are

---

17 Winner, L, 'Cyberlibertarian Myths and the Prospects for Community', 1997. Available at *www.rpi.edu/~winner/Cyberlib2.html*.
18 Jones, R, 'A Critique of Barlow's "A Declaration of Independence of Cyberspace" ', 1996, *Extropy#17* Vol.8(2).
19 Reidenberg, J, 'Governing Networks and Rule-Making in Cyberspace' 1996, 45 *Emory Law Journal* 911; Reidenberg, J, 'Lex Informatica: The Formation of Information Policy Rules Through Technology', 1998, 76 Tex L Rev 553. Reidenberg's *Lex Informatica* greatly influenced the work of Lawrence Lessig (discussed below), which now forms the focal point of the cyber-paternalist school.
20 On the role of software designers in default rule-making see Quintas, P, 'Software by Design', 1998 in Mansell, R and Silverstone, R (eds) *Communication by Design: The Politics of Information and Communication Technologies*, Oxford: OUP.

intersecting. He contended that in the light of *Lex Informatica*'s dependence on design choices, the attributes of public oversight associated with regulatory regimes could be maintained by shifting the focus of government actions away from direct regulation of cyberspace, towards influencing changes to its architecture.

The emergence of the cyber-libertarian/cyberpaternalist debate in the summer of 1996 led to a symposium on cyberlaw being held at the University of Chicago. During this symposium the distinguished law and economics theorist Frank Easterbrook made a powerful and challenging presentation.[21] His paper, entitled *Cyberspace and the Law of the Horse*,[22] presented to the assembled audience of cyberlawyers, stated that the subject did not exist. There was, he informed them, no more a 'Law of Cyberspace' than there was a 'Law of the Horse'.[23] Easterbrook charged the audience as being dilettantes – meddling in areas they didn't understand. He insisted that the best way to learn the law applicable to specialised endeavours, such as the internet, was to study general principles such as property, tort and contract. This gives rise to the question of whether Easterbrook is right. Instead of reading (and writing) textbooks on cyberlaw and cyber-regulation should we all be studying contract law, torts, property, intellectual property and criminal law? Bluntly, does cyberlaw exist as an extant subject worthy of academic study?

Cyberlaw and the challenge of the 'Law of the Horse'

Easterbrook's paper ignited an impassioned debate on the academic validity of the subject of cyberlaw and the role of cyberlawyers. His charge was quite different from that of the cyber-libertarian school. He did not claim, nor did not wish to claim, that cyberspace was to be treated as a separate and sovereign space due to the nature of 'bits': rather his indictment was that in debating the regulability of this place we had gone beyond the primary question of whether it was a suitable subject of specific regulation in the first place. Powerful advocates of cyberlaw responded immediately in an attempt to refute Professor Easterbrook's challenge. The most compelling response was that of Professor Lawrence Lessig in his Harvard Law Review commentary *The Law of the Horse: What Cyberlaw Might Teach*.[24] This paper was

---

21 Frank Easterbrook is/was Professor of Law and Economics at the University of Chicago, and is a Judge of the Court of Appeals for the 7th Circuit.

22 Published as Easterbrook, FH, 'Cyberspace and the Law of the Horse', 1996, U Chi Legal F 207.

23 This reference is made with regard to a statement by Gerald Casper, past Dean of Chicago Law School, who as Dean boasted that the School did not offer a course in The Law of the Horse.

24 113 Harv LR 501 (1999).

somewhat of a tour de force and had the twin effect of announcing the arrival of Professor Lessig as the leading commentator in cyber-regulatory theory, a position cemented later that year by the publication of his extended essay *Code and Other Laws of Cyberspace*, and in the minds of most cyber-regulatory theorists, satisfactorily answering the challenges set by Professor Easterbrook. An analysis of Lessig's response to Easterbrook though, leaves one a little unsettled. In his reply Professor Lessig chose to answer only one of Easterbrook's challenges and in doing so he believed he had adequately dealt with the wider question of the academic significance of the subject. Others, as we shall see, were not convinced.

The challenge Lessig answered was the narrow one of 'What is the value of a cyberlaw course?' Professor Easterbrook in his address had claimed that law schools should only teach courses that 'could illuminate the entire law' and that 'the best way to learn the law applicable to specialised endeavours is to study general rules'. It was on this point that Lessig took the fight to Easterbrook. Professor Lessig countered that unlike Professor Easterbrook, he believed that 'there is an important general point that comes from thinking in particular about how law and Cyberspace connect', that 'by working through examples of law interacting with Cyberspace, we will throw into relief a set of general questions about law's regulation outside of Cyberspace'. Throughout the rest of his commentary Lessig uses examples to illustrate regulatory tensions that arise when real space and cyberspace interface,[25] and then introduces his now famous concept of regulatory modalities and their effects both within and without cyberspace.[26] Professor Lessig's response to Easterbrook is simple and eloquent. While some specialised subjects, such as the Law of the Horse, are without academic merit due to their inability to shed light onto the study of the law at large, there are a few specialised subjects that merit study due to their ability to throw the general principles into relief, and cyberlaw qualifies as such. The force of conviction contained in Lessig's commentary, coupled with the powerful modalities of regulation analysis contained therein seemed to convince many commentators that Professor Lessig had adequately dealt with Professor Easterbrook's charge. Looked at more fully though, it becomes clear that Lessig has not rebutted the key indictments in Easterbrook's challenge to the cyberlaw community; instead, he has simply pled 'special circumstances'. By demonstrating that the cyber-regulatory community can give something back to the general legal debate, Lessig has bought some time for cyberlawyers. He had demonstrated why cyberlawyers may make a contribution to the study of the law, but he had failed to demonstrate why cyberlaw should not be seen as 'multidisciplinary dilettantism'.[27]

25 The examples used by Lessig are zoning speech, minor access and privacy protection.
26 Lessig's modalities of regulation thesis will be discussed in greater detail in Chapter 2.
27 Easterbrook, see above, fn 22 at p 207.

Were it not for the obvious power of Lessig's modalities of regulation thesis, he may have found himself held up as the archetypical model of Easterbrook dilettantism. Professor Lessig's thesis was predicated on the value regulatory theorists at large could draw from his new theory. In effect he was saying that effects in cyberspace demonstrate challenges from which regulatory theorists may learn. A lesser theory would instantly have left Lessig open to being challenged with committing a 'cross-sterilisation of ideas', Easterbrook's label for the 'put[ting] together of two fields about which you know little and get the worst of both worlds'. Even though Professor Lessig escaped such a fate he still hadn't truly demonstrated a sound academic foundation for the study of cyberlaw. Many specialised law courses could make unique contributions to the general study of law and legal or regulatory theory, but of itself this does not provide a foundation for them to be considered academic subjects in their own right. 'Space law' for instance has much to offer the law of property by contributing to the analysis of the limits of property and ownership and about trespass and access rights. This does not mean space law should necessarily be recognised as an academic subject.[28] It should therefore be no surpise that despite Professor Lessig's powerful defence of cyberlaw a variation of Professor Easterbrook's challenge was again delivered to the community of cyberlawyers in 2000 by Joseph Sommer, a lawyer in practice with the Federal Reserve Bank. Mr Sommer's paper *Against Cyberlaw*,[29] although not as sophisticated as Professor Easterbrook's, raised a new set of challenges. In essence the challenge of Mr Sommer was that social phenomena, not technologies, define laws and subjects of legal study. He demonstrates his thesis by pointing to the lack of longevity in any technologically defined set of laws. He reminds us that Abraham Lincoln was a 'railroad lawyer', a long-forgotten facet of the law, that there never was a law of the steam engine despite the huge social changes it brought about, and that much nineteenth-century race law could have been listed under a course in Cotton Gin Law, that being the technology which drove much of the development of race law and the law of slavery at that time. These are examples of the failure to develop technologically mediated laws. They demonstrate why 'neither Cyberlaw nor the law of the Internet exists, neither *can* exist . . . because technology and law are socially mediated, bodies of law do not respect technological boundaries, and technologies do not define law . . . Just as librarians do not classify books by their associated colour, lawyers should not classify fields of law by their associated technologies'.[30] Thus by the end of 2000, cyberlawyers actually faced two distinct attacks on the

---

28 The author would like to stress that equally he does not claim space law is not of academic merit.
29 15 *Berkeley Technology Law Journal* 1145 (2000).
30 This quote is made up from two separate quotes taken from *Against Cyberlaw* (above fn 29), one on p 1151 and the other on p 1153. The emphasis is in the original.

academic validity of their subject. The first, set in 1996 by Professor Easterbrook, was that cyberlaw created a risk of multidisciplinary dilettantism or
cross-sterilisation of ideas. This claimed that the subject was no more than a
fusing of two disparate subject-matters in the worst possible fashion.
Although this attack had been met by Professor Lessig, his defence was
incomplete. He had not managed to convince the academic jury that cyberlaw
was innocent of Professor Easterbrook's charges, although he had made a
sterling plea in mitigation, which had clearly bought the subject a degree of
academic recognition. The second charge, Mr Sommer's, remained unchal-
lenged. This is probably due to the fact that the burgeoning literature in
cyber-social theory was deemed to deal with his challenge without the need to
make a direct rebuttal.[31] The question still remains though – does cyberlaw
exist as an identifiable subject of academic study?

Cyber-society and computer science: Cyberlaw's contribution

By the summer of 1997 I had graduated from law school and had found my
first academic job. I was now a lecturer (assistant professor) in law at the
University of Stirling in Scotland. I was teaching intellectual property law,
contract and international trade law, but I had not been able to shake off that
excitement I had felt as an undergraduate in the spring of 1994. I wanted to
develop a course in information technology law, but I didn't really know
where to start as such courses were rare in the UK at the time. I decided I
needed to understand the technology better before I could develop my new
course. To this end I joined several mailing lists, mostly those organised by the
Internet Engineering Task Force (IETF) – of which more later. As with many
such subscribers I intended to simply lurk: my idea was to immerse myself
in the technical culture of the engineering community without revealing
myself as being an ill-informed bystander. Throughout that summer though,
list members kept discussing developments in the TCP/IP protocol. I knew
what TCP/IP was: transmission control protocol/internet protocol is the
dual communications protocol that manages and routes all packets of information within the network. I also roughly understood how it worked, but the
detail continued to elude me. Therefore, after some weeks of lurking I sent
out a request asking for an explanation of how these protocols functioned
and in so doing I revealed myself as an interested, non-technical, lurker. I
received one or two replies telling me to get off the list and to seek out lists
more suited to my level of knowledge: these thankfully were in the minority. I

31 Eg Castells, M, *The Internet Galaxy*, 2001, Oxford: OUP; Webster, F, *Theories of the Information Society*, 2nd edn, 2002, London: Routledge; Mansell, R, *Inside the Communication
Revolution: Evolving Patterns of Social and Technical Interaction*, 2002 Oxford: OUP.

also received several replies telling me where I could find the information I needed. Among these was a reply from Vint Cerf. He went into great length explaining how TCP/IP worked, its value to the network, and then he explained how he and Bob Kahn had developed these protocols between 1973 and 1978.[32] By throwing my question out into the community I had received a reply from one of the inventors of the protocol. Vint Cerf had never met me and I'm sure he had other demands on his time, yet he had taken the time to write quite a detailed email to me. This demonstrates the power of community within cyberspace. I believe this community provides the foundation of the answer to Joseph Sommer's social mediation argument.

Sommer's charge against cyberlaw is carried out on two levels. The first, which I have referred to above, is that technological practices seldom provide the framework for bodies of law. Thus the law of the steam engine, railroad and cotton mill have a short lifespan, as it seems will the law of space and of nuclear power. Some technologies do, though, determine new legal processes. Sommer himself discusses oil and gas law and to this list we may add aviation law, telecommunications law, copyright (or the law of the printing press) and media law. Why cannot cyberlaw (or internet law) be added to this illustrious list? Sommer seems determined that it cannot. His reason is that, 'only if we consider the internet to be a singular social phenomenon can we expect to see a law of the internet . . . however the internet is far too protean to support only a single set of social practices'. This is the application of Sommer's second-level attack on cyberlaw, his social mediation argument. This is based in the belief that as both law and technology are social endeavours, each must be understood through the lens of social interpretation (see Figure 1.1).

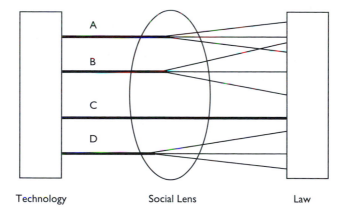

Technology   Social Lens   Law

Figure 1.1 The Social Lens.

32 A full discussion of the development of the TCP/IP protocol may be found in Hafner and Lyon, see above, fn 2 at 223–37.

When most technologies, such as the steam engine, pass through the lens, they are split into their component social effects (such as planning laws, contract, torts, trade laws and criminal laws). Only those pure technologies – such as technology C in Figure 1.1 pass though the lens unhindered and become a bona fide body of law in their own right. Thus oil and gas law is recognised as it is of a singular social phenomena, whereas more socially complex technologies, such as the railroad, are split into components. Almost immediately though Sommer runs into problems with his thesis. He must deal with bodies of law deriving from such complex technologies. He dismisses the 'law of the telephone', or telecommunications law as 'a specialised body of regulatory law, of little interest outside the industry', while copyright law is dismissed as 'mediated through the first amendment . . . not a law of the printing press'. These are quite inadequate responses to the complex bodies of rights that surround these technologies. While there is no doubt that much of the foundations of telecommunication law has a basis in industry regulation,[33] it is a much more socially rich body of law than Sommer would suggest. These regulatory regimes ensure vital social provisions such as universal access, interoperability, consumer protection and price controls. There is a socio-legal tradition in the study and teaching of telecommunications law,[34] and although specialised there is no doubt it is a recognised niche legal subject. Similar defences may be made of aviation law, media law and arguably even space law. Each are based on strong theoretical foundations and are steeped in the socio-legal tradition. They are socially complex, yet recognisable legal disciplines. Perhaps the weakest provision of Sommer's argument is that copyright law, a long-established and respected course of study, has minimal links with its technological foundations. Sommer claims copyright has 'always been mediated through the first amendment'. This reflects a very US-centric view of the discipline of copyright. Copyright developed in the UK, not the US.[35] In fact it took over 100 years for the US to recognise the copyrights of other nations and when it eventually did so this was not provided for through the first amendment, but was merely an extension of

---

33 Walden, I, 'Telecommunications Law and Regulation: An Introduction' 2001, in Walden, I and Angel, J (eds) *Telecommunications Law*, Oxford: Blackstone Press.
34 For example the Centre for Socio-Legal Studies at Wolfson College, Oxford offers a programme in Comparative Media Law and Policy, which has a telecommunications element. Similarly, the London School of Economics offers an MSc in Media and Communications Regulation and Policy, which is co-taught by the Law Department and the Department of Media and Communications. This programme has a high socio-legal content.
35 Birrell, A, *Seven Lectures on the Law and History of Copyright in Books*, 1899, London: Cassell & Co; Garnett, K and Davies, G (eds), *Copinger and Skone James on Copyright*, 14th edn, 2002, London: Sweet & Maxwell.

already existing protectionist provisions.[36] Prior to the enactment of the Human Rights Act in 2000, the UK never had a clearly defined 'right to free expression'; therefore UK copyright law could not be mediated through such a right. A study of the history of UK copyright law reveals that copyright developed from 'Letters Patent', which were promulgated by successive monarchs specifically to regulate the challenges posed by the development of the moveable type printing press.[37] Many competing social interests were challenged by this invention: the churches' control of religious texts, the state's management of political and satirical texts, the monopoly over reproduction rights held by the Stationer's Guilds, and the rights of authors, playwrights and poets. In many ways the printing press caused a similar social upheaval to the internet. Copyright was the legal response to this. The printing press is only the starting point of a history that demonstrates a symbiotic relationship between technology and copyright law. As technologies for exploiting creativity developed, copyright law evolved. First, the development of the photographic plate and camera in the 1840s saw, with the evolution of photography, the extension of copyright to photographic images by the Fine Art Copyright Act 1862. This was followed by several extensions to copyright to recognise, and regulate, the new technologies of sound recording,[38] audio and video broadcast by radio waves,[39] film (or moving picture) recording[40] and cable distribution.[41] To this day copyright, with the extension of copyright

36 Copyright was extended to American authors by the Copyright Act of 1790. It was not until the Chase Act of 1891 that the US extended copyright protection to authors of selected countries on a bilateral basis. For an analysis of the background to this period in US Copyright history see Post, D, *Some Thoughts on the Political Economy of Intellectual Property: A Brief Look at the International Copyright Relations of the United States*, 1998, paper presented to the National Bureau of Asian Research Conference on Intellectual Property, Chongqing, China, September 1998, available from: *www.temple.edu/lawschool/dpost/Chinapaper.html*.

37 In the 200 years following the development of the printing press successive monarchs used Letters Patent to control the publishing industry. The first of these is probably that of Henry VII, who in 1504 appointed William Facques as the first royal printer, granting him the exclusive right to print official documents. Several such grants and orders followed including Henry VIII's order of 16 November 1538, which provided that 'all new books had to be approved by the Privy Council before publication'. Most famous among these Letters Patent is the *Statute of Mary*, 1557 which 'granted a royal charter that limited most printing to members of the Stationers' Company, and empowered the stationers to search out and destroy unlawful books'. Many commentators now see the *Statute of Mary* as the precursor of the *Statute of Anne*, 1709, which is recognised as the world's earliest Copyright Act. For a discussion of the history of English Copyright see Patry, W, *Copyright Law and Practice*, 7th edn, 1996, Washington DC: BNA Books, ch 1.

38 First protected by the Copyright Act 1911.

39 First protected by the Copyright Act 1956.

40 Films were first protected as a series of individual photo images. It was only latterly with the Copyright, Designs and Patents Act 1988 that films were given distinct protection in the UK.

41 First protected by the Cable and Broadcasting Act 1984.

into digital media, remains mediated by technological developments to a far greater degree than rights of expression.[42]

If it is not the social lens that determines which technological changes merit a body of law, what does? I believe that Sommer was arguing along the correct lines, but that he was attracted by the simpler explanation. He is correct that the social lens determines which technologies merit a body of law, but wrong to assume that it is only 'singular social phenomena' that benefit. The actual position is that some technologies have an identifiable socio-legal effect beyond their direct contribution to the fabric of society, while others do not. Thus the aviation industry created a complex set of socio-legal requirements for entry and exit from national airspace, overflying rights, regulation of supersonic travel, emissions, language provisions (English is the international language of the air) and much more. Thus the development of passenger flights had an impact on nearly every corner of law and society, an impact which rather unusually the rail industry did not have. The rail industry was more of a standard transport industry. It promulgated immense social change (the UK did not have a standardised time until Railway Time was introduced), yet had little legal impact. The current legal regimes could cope with almost all changes introduced by railways. The best way to illustrate this effect is in Figure 1.2. Here we see two technologies of powerful social impact (technologies A and D), which are of little or no legal impact. Their changes are completely societal with no legal/regulatory effect: these are technologies such as the railroad or the steam engine. Then we have

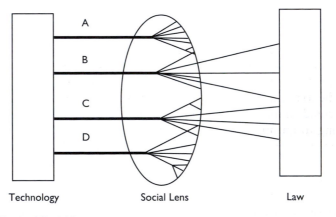

Figure 1.2 Revised Social Lens.

42 As can be seen with the promulgation of the Digital Millennium Copyright Act 1998 (USA) and the Directive on Copyright and Related Rights in the Information Society (EU). Both were promulgated to protect digital content management systems in the face of assaults from file-sharing technologies.

technologies B and C. They are also of immense social impact, but some aspects of this technology goes beyond the social and are also of legal/regulatory effect. These are technologies such as commercial aviation, print and broadcast media, and telecommunications. This begs the question, where does the internet fit into this model?

I believe the answer to that question is to be found in my earlier example of internet culture. The technologies which are of socio-legal, as opposed to just social, effect are those that create new social challenges requiring a legal/regulatory response. The internet is such a technology: Lawrence Lessig has already demonstrated as much in his paper *The Law of the Horse: What Cyberlaw Might Teach*.[43] Further, I believe that the fact that cyberspace creates new communities, communities where junior law lecturers can discuss technical matters with senior technicians six thousand miles away, demonstrates the socio-legal challenge of the internet. If we examine which technologies give rise to socio-legal implications, the list is overwhelmingly populated with those technologies that gave rise to new communities linked through the medium of the technology: the printing press, radio and television broadcasting technologies, telecommunications and aviation. Where technology fails to create communities, but merely serves them (the steam engine, the cotton gin and arguably the rail network) it never breaks out of the social lens. Cybercommunities are unique and centred around cyberspace. They are creating socio-legal demands. On this basis, cyberlaw passes the refined version of the Sommer social mediation test.

The Easterbrook challenge remains. Professor Easterbrook's argument against cyberlaw is at the same time more simple and yet more complex than Sommer's. The simplicity of Easterbrook's argument is in its construction: the study of specialised subjects such as cyberlaw or the law of the horse fail to illuminate the general body of legal knowledge and are therefore dilettantism. The complexity is in dealing with what is basically an opinionated standpoint. Easterbrook believes that 'beliefs lawyers hold about computers and predictions they make about new technology, are highly likely to be false'. I could simply say I disagree, but this fails to rebut the principle. By responding to an opinionated standpoint with a counter-opinion one fails to advance the debate: this is the complexity of Easterbrook's position. To counter the Easterbrook challenge therefore, one must engage with his analysis at a more basic level in the hope of challenging the foundations upon which he bases these opinions. To do so I'm going to turn Easterbrook's argument on its head and ask what is the value of a course in law and economics? The Law and Economics School was developed at the University of Chicago and is founded upon Ronald Coase's Nobel Prize winning theorem that 'in the absence of transaction costs, all government allocations

43 See above, fn 24.

of property are equally efficient, because interested parties will bargain privately to correct any externality'. This is set out in his famous 1960 paper *The Problem of Social Cost*,[44] in which he proposed that the role of law, and the role of lawyers, was to encourage market efficiency by removing or limiting transaction costs and externalities wherever possible. Professor Easterbrook is of course one of the leading lights of the law and economics movement. At Chicago Law School he, along with others such as Professor Gary Becker and Judge Richard A Posner, have schooled the legal world in Chicago supply-side economics. Such is Easterbrook's reputation for market efficiency arguments that Lawrence Lessig tells us that one year in his cyberlaw class a student created an online character called FEasterbrook, who would substitute the word fair with efficient in all conversations with him, thus 'this is not fair' became 'You mean it's not efficient'.[45]

What though does the Law and Economics School contribute to our understanding of the law in general? Easterbrook, and his colleagues, would no doubt point to several new approaches to contract, property, torts and criminal law, which have developed as part of the Chicago Law School contribution to the general understanding of law through the applied study of law and economics. He would, quite rightly, be extremely proud of this contribution. In the 1950s though, when the discipline of law and economics was in its infancy, how would he have reacted to a charge that 'beliefs lawyers hold about economics and predictions they make about new developments, are highly likely to be false'? Certainly he would not have acquiesced. He would surely have pointed to the work of respected economists such as Ronald Coase who were happy to contribute to this burgeoning academic study. In fact he does as much in his 1996 address saying, 'The University of Chicago offered courses in law and economics, and law and literature, taught by people who could be appointed to the world's top economics and literature departments'. This is a vital structural support for Easterbrook's argument; it is also its weakness. Here he acknowledges, albeit implicitly, that law, as a social science, has a duty to listen to, and engage in discussion with, other social science subjects. He lists literature and economics as partner subjects but equally the list of 'acceptable' subjects could have included, sociology (law and society), anthropology, philosophy and political science. Why is law and society or law and anthropology acceptable, but law and equine studies (or the law of the horse) not? The answer is now clear: whereas anthropology and sociology are identifiable, and accepted, social science subjects, which may illuminate and influence our wider understanding of the law, understanding the mechanics of trading, training and running horses, will not offer the same contribution. Law must not be seen in isolation: it is part of a wider

44 Coase, R, 'The Problem of Social Cost', 1960, 3 J Law & Econ 1.
45 Lessig, see above, fn 10, p 12.

community of social sciences. What about cyberlaw? Unfortunately the attitude represented by Easterbrook is one that is all too common among 'classical' social scientists.[46]

Those who study and work with computers have long been discriminated against by the wider academic community. Computer science was seen for many years not to be a truly academic subject. In the 1950s computer scientists were dismissed as 'technicians' by those with a background in social and natural sciences. At MIT's Lincoln Laboratory, a leading computer research centre, only fully fledged physicists and mathematicians were allowed to join the research staff.[47] However, a group of computer scientists, led by Frank Heart, broke through this glass ceiling and by the late 1960s a degree of academic respectability had been obtained by early graduates in computer science, such as Vint Cerf, Steve Crocker and Jon Postel. Their efforts saw computing science (or information systems) gradually establish itself as a recognised academic subject across college campuses in the US, and into the universities of Europe. By 1979 there were 120 academic computer science departments across the US alone.[48] It took nearly 30 years, but by 1980 computer science was accepted as a valid subject of academic inquiry. One of the reasons why it took so long to be recognised is that it straddles the boundary between the natural sciences and the social sciences. Some computer scientists look very much like physicists or mathematicians in their chosen field of study, while others look very much like sociologists or even lawyers. The Massachusetts Institute of Technology (MIT) – one of the world's leading institutions for the study of natural sciences – is home to one of the oldest and most respected computer science departments. At MIT the subject was historically approached as a development of the study of mathematics and physics.[49] Now, the world-leading Laboratory for Computer Science (LCS), provides an umbrella for researchers of diverse backgrounds. The LCS is home to 23 research groups, which support just over 500 staff,

---

46 I am using the term 'classical' here to denote social scientists who elevate the classical social science disciplines such as sociology, anthropology, economics, politics and history above 'modern' social sciences such as business studies, industrial relations, gender studies and computer science and information systems.

47 Hafner and Lyon, see above, fn 2, p 89.

48 This information is all drawn from Hafner and Lyon, see above, fn 2.

49 In the early 1950s MIT/Lincoln Labs worked jointly on two large-scale computing projects for the Defense Department. The first was Whirlwind, a computer project designed to manage and analyse communications from the Distant Early Warning radar array. This was followed by the Semi-Automatic Ground Environment (SAGE) project, which received, analysed and then acted upon tracking data from the DEW array. Most technicians working on Whirlwind/SAGE were physicists or mathematicians. It should be noted that MIT now takes a much more holistic view of the subject. For details on Whirlwind/SAGE see Hafner and Lyon, above, fn 2, at pp 30–1. For details on the history of the Laboratory for Computer Science at MIT see: *www.lcs.mit.edu/about/about.html*.

including approximately 100 faculty and research staff and 300 graduate students. The Laboratory's research falls into seven principal categories,[50] which encompasses all aspects of computers, technology and society. Its foundations as a centre of technological excellence may be found in its programmes in network design,[51] architecture[52] and software.[53] It has also broadened its horizons to examine the effects computers and information technology (C&IT) are having on society. Research programmes such as its theory programme[54] and its interfaces and applications programme,[55] as well as its participation in Project Oxygen, a project aimed at inventing and developing pervasive, human-centred computers, demonstrate that the Laboratory now takes a holistic approach to C&IT research. Similarly, the London School of Economics and Political Science (LSE), long-recognised as one of the leading social science institutions in the world,[56] has a flourishing Department of Information Systems.[57] Here, researchers carry out cutting edge social science research examining the effects of computers and digitisation on individual communities and society at large. C&IT research, in particular network theory, is now quite clearly accepted by the academic community as a valid research subject for social scientists. As a result, one cannot distinguish computer science from any of the other accepted bodies of social science research, such as economics, literature, sociology or politics. It therefore must be the case that lawyers and computer scientists can learn from each other through discourse and debate. Like the 'classical' social sciences listed, C&IT research is now recognised as one of the family of social sciences that may illuminate and influence our understanding of the law.

50 These are: networks and systems, architecture, software, theory, interfaces and applications, computer science and biology, and the World Wide Web.
51 This research encompasses issues in connection with mobile and context-aware networking and building and investigating high-performance, practical software systems for parallel and distributed environments.
52 This research investigates the development of architectural innovations by directly compiling applications onto programmable hardware, by providing software-controlled architecture for low energy, and easier hardware design and verification.
53 This research seeks to improve the performance, reliability, availability and security of computer software by improving the methods used to create such software.
54 This studies the theoretical underpinnings of computer science and information technology.
55 This develops technology that facilitates direct user interaction, including computer graphics and conversational interfaces using spoken dialogue, and to explore roles computer science can play in user- and patient-centred applications.
56 In the *Sunday Times University Guide for 2004*, it was noted that: '[the] LSE's claim to be the leading social science institution in the world is no hollow boast'.
57 This Department describes itself on its website as: 'one of the largest departments of its kind in the world. It is well known for its research and teaching in the social, political and economic dimensions of information and communications technology. It covers most areas of information systems and represents a range of academic approaches and specialisms, from systems design and management to theory and philosophy.'

As such, to fail to recognise cyberlaw would be as perverse as a failure to recognise the disciplines of law and economics, socio-legal studies, law and anthropology or jurisprudence. It would mean a failure to recognise law in its correct social context and would lead to gaps in our understanding of the law. We must therefore recognise 'Law and Computer Science' (or cyberlaw).[58] It is not, as Easterbrook claims, multidisciplinary dilettantism; it is essential for our illumination of the science of the law. As such we may rebut Easterbrook's charge.

The role of the cyberlawyer therefore seems to be assured. Despite strong challenges to both the academic validity and rigorousness of the subject, it can be demonstrated to be both legitimate and academically sound. Lawyers, regulators and computer scientists have much to learn from studying this subject. It is to be hoped they approach it with open eyes and an open mind.

58 Cyberlaw here is being used in its natural sense as the study of law and the internet. The author recognises this is different from 'law and computing science' or IT Law. The bulk of social science research carried out in IS or CS departments is, though, to do with networks and network effects. Thus within IS the focus of social science research is in internet research and thus a parallel with cyberlaw may be drawn without extending the definition of cyberlaw.

Chapter 2

Regulatory competition and webs of regulation

> *Regulations grow at the same rate as weeds.*
>
> *Norman R. Augustine*

War, wrote Clausewitz, is the continuation of politics by other means.[1] In this famous statement Clausewitz defines war as a political act – the political object being the first and highest consideration in the conduct of war. An alternative iteration of Clausewitz's famous dictum may be to say that war is the application of regulation through the use of force. By applying a wide definition of regulation as 'all forms of social control, state and non-state, intended and unintended',[2] we see that war is a logical (final) choice when regulators find themselves in competition with one another. When two competing regulatory authorities, A and B, attempt to impose their regulatory ideologies upon a group of persons (or upon a geographical place) competition for acceptance may occur at several levels. At a socioeconomic level, regulators may compete for recognition by offering economic or social incentives, thereby utilising the market to achieve a settlement. Alternatively, competing regulators may appeal directly to the public at large to safeguard their regulatory role by winning the support of a majority (or a sufficiently large minority) of the populous. This, of course, forms the heart of democratic

---

1 von Clausewitz, C, *On War*, 1832, Graham, J and Maude, F (trans), London: Routledge, 1968, p 23.
2 The concept of regulation may currently be narrowed down to three accepted theories or descriptions of the phenomenon. These are presented in the current literature of the field as (i) the presentation of rules and their subsequent enforcement usually by the state, (ii) any form of state intervention in the economic activity of social actors, or (iii) any form of social control whether initiated by a central actor such as the state or not and including all acts whether they are intended to be regulatory or not. In this book we will use the last of these definitions, sometimes referred to as 'the European approach' and often used as a synonym for governance. See Baldwin, R, Scott, C and Hood, C, *A Reader on Regulation*, 1998, Oxford: OUP at p 4; Hood, C, *Explaining Economic Policy Reversals*, 1994, Maidenhead: Open UP, p 19.

regulation by popular government. This technique is also used though by unelected regulators to gain the popular high ground over potential competitors.[3] In such situations we may say that they are seeking to obtain a social or community-based settlement. Finally, regulators may seek the protection of a more senior regulatory authority. By appealing to those with power to decide who should regulate and how, junior regulatory authorities seek protection from their senior partner. In effect this happens frequently: regulatory bodies often lobby government to safeguard their regulatory domain, and in many cases to seek to extend it further.[4] Such competition for recognition finds its basis in structures of control or hierarchies. Thus we can see that competition for recognition, capacity to regulate, and for finances and resources occurs frequently between regulators. The ultimate expression of competition for resources, recognition, power and finances is when two nations (competing regulators) go to war over an area of land or an issue of political or religious ideology. Thus war is the continuation of regulation by other means. This though is only the beginning of regulatory competition and the regulatory web.

Regulatory competition, regulatory modalities and systems

The key to understanding regulation is in understanding how regulatory conflicts arise and how they are resolved. The nature of the regulatory environment is such that regulation surrounds and controls our actions from the moment we are born to when we die. Some areas of our lives are clearly and noticeably regulated, often for our own protection or for the protection of others. Thus potentially hazardous activities such as driving, working with heavy plant or machinery, or playing contact sports are closely regulated through a network of licences, professional bodies and professional standards. Some areas of our lives are regulated for the good of our fellow citizens. There are tight regulations covering for example air travel, television and media, and antisocial activities such as excess consumption of alcohol or excess noise. In addition to such obviously regulated activities, all aspects of our everyday activities such as walking to work or shopping in a supermarket are equally closely regulated, but in a less formal manner. You may, for example feel it would be a wonderful idea to walk to work naked on a particularly hot summer day, but a collection of social and legal controls

---

3 For example on 3 September 2003 the Strategic Rail Authority took out a full-page advert in the London *Times* at a cost of £28,000 to rebut a story published the previous evening in the London *Evening Standard*.

4 A good example of such a campaign is the AURE (Alliance of UK Health Regulators on Europe) lobby campaign against the draft EU Directive on Mutual Recognition of Professional Qualifications.

would ensure that you did not get too far from your front door. Equally, you may feel there is nothing wrong with opening packets in the supermarket to taste items before deciding whether or not to purchase them, but again a network of social and legal controls will regulate your actions. Finally, there are some areas of our lives where to say we are controlled or regulated may, at first instance, seem fanciful, but which are in fact equally closely controlled. As humans we have wishes and desires, some of which we hope we are destined to fulfil, but some of which we know from the rules of this place will remain unfulfilled. We may idly wish for many things from immortality to invisibility to the ability to travel instantly without form. These things are, though, impossible (at least as we currently understand the universe), thus even though there is no man-made regulation that prevents any of these things, the environment regulates and controls us as effectively as any law or norm.

These examples illustrate the natural order of the regulatory environment. In any situation there will be an extant, external, regulatory settlement that controls the actions, needs or desires of the regulatee. There will also be an internal value set to which the regulatee will refer when making any decision. The regulatee, by reference to this value set, may choose to accept, or to challenge the regulatory settlement. This decision effects the actions of both the regulatee and the extant regulatory settlement. For example in 2003 ex-Royal Marine Stephen Gough made the 900-mile journey from Land's End to John O'Groats in the nude, earning himself the soubriquet *The Naked Rambler*, a journey he set out to repeat in April 2005 and which to date he has yet to complete, due to the intervention of the authorities.[5] Mr. Gough has chosen to do this not simply out of a desire to do something different; he is doing it to challenge the current regulatory settlement that casts the naked human body as an indecent object that requires to be clothed.[6] Unsurprisingly, Mr. Gough's actions have attracted the attention of the regulators and he has been arrested and charged several times by the authorities, including his current detention in Edinburgh.[7] His actions have also successfully attracted a degree of publicity for his cause and some minor debate has taken place in the media about his campaign.[8] Unfortunately for Mr. Gough though, his challenge to the accepted regulatory settlement seems to be doomed to fail as a result of a high degree of public indifference to his

---

5   At the time of writing (April 2006) Mr. Gough is in prison in Edinburgh sentenced to three months imprisonment for contempt of court.
6   BBC News, *Naked Rambler Completes His Trek*, 22 January 2004. Available at: *http://news. bbc.co.uk/1/hi/scotland/3420685.stm*.
7   For the full story see *Edinburgh Evening News, Naked Rambler Exposed to Jail Term by Sheriff*, 7 April 2006. Available at: *http://news.scotsman.com/topics.cfm?tid=952&id= 534562006*.
8   Full details may be accessed via the Naked Walk website at *http://nakedwalk.org/*.

campaign and the existence of a settled and clear set of publicly supported regulations on this matter. Not all challenges to the extant regulatory settlement are, though, equally impotent and regulatory flux is common. An example of a successful challenge to accepted regulatory standards may be found in the successful campaign pursued by the UK anti-hunting lobby against the hunting of foxes with hounds. Hunting with hounds has a long history in the UK with the earliest recorded instances taking place in sixteenth-century Norfolk. The practice quickly spread and by the end of the seventeenth century there were organised packs hunting both hare and fox across the UK. By 2005 there were an estimated 200,000 people involved in fox hunting in the UK. But despite its long history and wide geographical extent, fox hunting has long been seen by a substantial portion of UK society to be a socially objectionable, even barbaric, activity. For over 50 years anti-hunt protesters pursued a campaign to ban fox hunting,[9] a campaign that spanned the spectrum from peaceful protests such as lobbying, petitions, sit-ins and picketing to violent protest including hunt sabotage and which finally came to fruition with the passing of the Hunting Act 2004. The Act, which came into force on 18 February 2005, finally banned the hunting of any wild mammal using dogs, overturning a regulatory settlement which was at least 400 years old.[10] In this case members of the anti-hunt lobby, by following their internal compass rather than the external regulatory settlement, eventually brought about a change in that regulatory settlement.

These two simple case studies serve to illustrate the complexity of the modern regulatory environment. Any attempt to map even a simple regulatory settlement involves recording the current regulatory status quo, allowing for regulatory flux caused by external (technological, economic or environmental) changes and accounting for any changes in social values through internal (moral) evaluation of any such regulation. Faced with such complexity it appears that any attempt to model regulation in modern society is impossible. In an attempt to deal with this problem regulatory theorists turned to systems theory. Systems theory was first proposed in the 1940s by the biologist Ludwig von Bertalanffy as an effective methodology to model complex systems.[11] Von Bertalanffy emphasised that real systems are open to, and interact with, their environments, and that they can acquire qualitatively new properties through such interaction, resulting in continual evolution. Rather than reducing an entity, such as the human body, to the properties of its individual parts or elements such as cells, systems theory focuses on the

---

9 A full history of the campaign may be found at: BBC News, *Timeline: Hunting Row*, 17 February, 2005. Available at: *http://news.bbc.co.uk/1/hi/uk_politics/1846577.stm*.
10 Section 1 of the Hunting Act states: 'A person commits an offence if he hunts a wild mammal with a dog, unless his hunting is exempt'.
11 For a discussion see von Bertalanffy, L, *General System Theory: Foundations, Development, Applications*, 1968, Harmondsworth: Penguin.

arrangement of and relations between the parts that connect them into a whole. This determines a system that is independent of the substance of the elements. Thus, the same concepts and principles of organisation may be used to underlie several distinct disciplines (physics, biology, law, sociology, etc.), providing a basis for their unification. A prime example of a complex system is illustrated in the gardener's dilemma. The gardener's dilemma asks us to imagine a garden of many different plants from many different species, all managed by a gardener who seeks to maximise some variable over the garden as a whole.[12] The garden has some general characteristics. First, individual plants are heterogeneous: the relationship between the individual plant's state (for example pruned or unpruned, fed or unfed) and its development is different for each plant. Second, there are substantial external effects between individual plants. This means the development of each plant can be effected, positively or negatively, by the condition and development of other plants. Third, each plant's response to being in one state or another (for example fed or unfed) is endogenously determined, that is it is a function of the state of some number of other plants. For example a plant's response to being fed may depend on whether it is receiving sufficient sunlight; this in turn may be determined by whether the plant is surrounded by pruned or unpruned neighbouring plants. The gardener's dilemma is to identify the system configuration for the garden, which produces the maximum yield for the garden as a whole. You might imagine that solving the gardener's dilemma would not prove too difficult. You may begin by mapping the individual characteristics of each plant – pruning and/or feeding plants in isolation. Next you may move plants in different stages of development around the garden to measure the external effects of other plants upon the test subjects. But, in so doing you effect the state of the other plants in the garden, as they are dependent on the state of your test plant (external effects). The changes you made to your test plant altered the parameters of the second test. It is in fact impossible to solve the first test without solving the second and vice versa. The problems are recursive; each possible configuration of the system, and there may be a massive number of configurations,[13] presents a different problem to be solved. The system is 'caught in a web of conflicting constraints [in which] each element of the system affects other parts of the whole system and changing [a single element] will have affects which ripple out throughout the system'.[14] This is the ripple effect noted by Lon Fuller and

---

12 The following section is based upon and draws heavily from Post, D and Johnson, D, 'Chaos Prevailing on Every Continent: Towards a New Theory of Decentralized Decision-making in Complex Systems', 1998, 73 Chi-Kent L Rev 1055.

13 In a system with N elements, each of which may take S possible states, there are S^N possible system configurations. A ten-plant garden in which each plant could be in four possible states has $4^{10} = 1,048,576$ different configurations.

14 Kauffman, S, *At Home in the Universe*, 1995, Oxford: OUP, p 173.

which he labelled 'polycentric'.[15] You may assume, that the gardener's dilemma, and Fuller's polycentric web, may be mathematically modelled, and that solutions may be devised. This though is not the case. There is an abundant literature, primarily in the fields of economics and computer science, which demonstrate that the problems characterised by the gardener's dilemma are 'computationally intractable': effectively they are insoluble by known techniques because the number of discrete computational steps required to calculate the solution increases exponentially with the size of the problem.[16]

Systems theory's main effect on regulatory theory was to suggest the 'law of requisite variety', which was developed by psychiatrist and cyberneticist W. Ross Ashby in his *Introduction to Cybernetics*.[17] This states that 'in active regulation only variety can destroy variety'.[18] This rule suggests that when modelling control and accountability in a complex system it is impossible to know with any precision what the effect of any change in the regulatory status quo might be without first mapping all the potential points of disturbance caused by the change and accounting for them in the control system.[19] But as a complex system is computationally intractable this is impossible, meaning that a regulator can never be sure, in any complex system, what effect his actions will have. Ashby's law is particularly disturbing for regulators and regulatory theorists alike. For regulators, it suggests they can never be sure what effect their intervention into the regulatory status quo will cause: it may achieve the desired result, or much like the gardener pruning a bush, it may have unforeseen consequences. For regulatory theorists it suggests that any attempt to map regulatory systems would be futile. Rather than be downhearted by Ashby's law, regulatory theorists have instead sought to draw on systems theory by seeking to identify and evaluate those areas of commonality that provide connections between the different components of the system. Thus the approach most commonly used today is to identify

15 See *Modelling Regulation in Cyberspace: Polycentric Webs and Layers* below, and Fuller, L, 'The Forms and Limits of Adjudication', 1978, 92 Harv L Rev 353.

16 An excellent overview of the development of impossibility theorems may be found in Rust, J, *Dealing with the Complexity of Economic Calculations*, 1996, available at: *http://econwpa.wustl.edu:8089/eps/comp/papers/9610/9610002.pdf*.

17 Ashby, WR, *An Introduction to Cybernetics*, 1956, London: Chapman & Hall. See in particular ch 13 'Regulating the Very Large System'.

18 Ashby, WR, 'Variety, Constraint, and the Law of Requisite Variety.' in Buckley, W (ed) *Modern Systems Research for the Behavioral Scientist*, 1968, Chicago: Aldine.

19 In Ashby's Law each disturbance D will have to be compensated by an appropriate counter-action from the regulator R. If R would react in the same way to two different disturbances, then the result would be two different values for the essential variables, and thus imperfect regulation. This means that if we wish to completely block the effect of D, the regulator must be able to produce at least as many counteractions as there are disturbances in D. Therefore, the variety of R must be at least as great as the variety of D.

and evaluate macro-regulatory modalities that may be used by regulators to control patterns of behaviour within complex systems. A result of this practice may be seen in Robert Baldwin and Martin Cave's book *Understanding Regulation*.[20] In this the authors outline eight (alternative) regulatory strategies: (1) command and control, (2) self-regulation, (3) incentives, (4) market-harnessing controls, (5) disclosure, (6) direct action, (7) rights and liabilities laws, and (8) public compensation. The authors describe these eight strategies as the application of the 'basic capacities or resources that governments possess and which can be used to influence industrial, economic or social activity'.[21] Thus government may (a) use legal authority and the command of law to pursue policy objectives, or it may (b) deploy wealth through contracts, loans, grants, subsidies or other incentives to influence conduct, or (c) harness markets by channelling competitive forces to particular ends, or (d) deploy information strategically, or (e) act directly by taking physical action, or (f) confer protection to create incentives. Thus Baldwin and Cave examine how government may influence the outcome of any situation by applying a mixture of incentives and controls to achieve the desired outcome. This removes the discussion from the detail of the regulatory settlement, where the Law of Requisite Variety suggests analysis would be futile, to the higher level of regulatory structures and strategies.

This traditional model can also be seen in Mark Thatcher's summary of regulatory interpretations.[22] In attempting to provide a model that simplifies the variety of regulatory strategies discussed above, Thatcher models how regulators seek to influence behaviour and suggests four families of regulatory interpretation: (1) *classical economics*, where regulation is an interference in the market that may be necessary, (2) *political economy*, where regulation is inherent to society, and is used by the state to ensure that the market functions, (3) *political science and law*, where regulation steers public activity and is concerned with controls over private activity, and (4) *sociological*, where regulation is achieved through informational norms that guide behaviour. An attempt to extend the traditional model of regulatory analysis into cyberspace was made by Lawrence Lessig in his extended essay *Code and Other Laws of Cyberspace*.[23] In this Lessig seeks to identify four 'modalities of regulation': (1) law, (2) market, (3) architecture and (4) norms, which may be used individually or collectively either directly or indirectly by regulators. To

20 Oxford: OUP, 1999.
21 Ibid at p 34. The basic capacities and resources referred to are: (1) to command, (2) to deploy wealth, (3) to harness markets, (4) to inform, (5) to act directly and (6) to confer protected rights.
22 Thatcher, M, 'Explaining Regulation Day 1 Sessions 4 and 5', 2000, a paper delivered at the Short Course on Regulation, The London School of Economics and Political Science, 11–15 September 2000.
23 New York: Basic Books, 1999.

explain his concept Lessig uses a simple example to demonstrate how his four modalities may be used by regulators. In seeking to regulate your decision whether to smoke or not, a regulatory body may promulgate laws to constrain the supply of cigarettes, such as age restrictions, or to regulate your ability to consume cigarettes in certain environments, as with the recent outright ban against smoking in designated public places in Scotland.[24] Alternatively, regulators may encourage the development of a certain standard of norms such as media campaigns, which demonstrate the dangers of passive smoking, or those designed to paint smokers as generally antisocial. These are designed to encourage a strong, and negative, societal response to smoking, particularly smoking in public places. Equally, a market solution may be used, such as in the UK where a robust policy of elevated tobacco duties is used to discourage smoking. Finally, architectural solutions may be used: cigarettes with filters may encourage more smoking as they are perceived as being less dangerous while nicotine-enhanced cigarettes will prove to be more addictive. Each modality thus has a role to play in regulating your decision.[25] Lessig suggests that the true regulatory picture is one in which all four modalities are considered together. Regulators will design hybrid regulatory models choosing the best mix of the four to achieve the desired outcome.[26] Similarly, Colin Scott and myself in our paper *Controlling the New Media*[27] suggest a focus on hybrid models of regulation. We, like Lessig, suggest four modalities of regulation which we title, (1) hierarchical control, (2) competition-based control, (3) community-based control and (4) design-based control. Unlike Lessig, we acknowledge that the development of regulatory structures is often organic in nature,[28] though we imagine regulatory bodies, through the employment of hierarchical controls, fashioning the structure of such organically developed systems.[29] Thus, ultimately in this paper we support the consensus that regulators design regulatory systems. Thus these models all share a common foundation. All are modelled upon the belief that regulatory designs are based upon active choices made by regulators: they suggest a regulator who works within a settled environment and who has time to positively consider policy decisions. It is as if all the variables in the garden in the Gardener's Dilemma have suddenly frozen in time, allowing the gardener to

---

24 Smoking, Health and Social Care (Scotland) Act 2005; The Prohibition of Smoking in Certain Premises (Scotland) Regulations 2006, SSI 2006/90.

25 Lessig, see above fn 23, p 87.

26 Ibid, pp 87–8.

27 Murray, A and Scott, C, 'Controlling the New Media: Hybrid Responses to New Forms of Power', 2002, 65 MLR 491.

28 Ibid, p 505.

29 At p 505 we say, 'we contend that an emphasis on hybrid forms of control will tend to lead to the deployment of hierarchical controls as instruments to steer organic or bottom-up developments'.

evaluate the best approach to take. Should s/he pursue a policy of pruning, or watering, to achieve the desired result? What about those occasions though when a regulator is called upon to act immediately? When a regulatory conflict occurs (that is when one modality of regulation is altered by social or environmental changes and negatively impacts upon settled regulatory practice) it is less likely that the regulator will have time to make a positive decision. What results, in terms of regulatory practice, will be the outcome of such regulatory competition?

Regulation, technology and social change

In the modern world regulatory competition occurs widely and arises rapidly. The greatest cause of such competition is rapid technological advancement. Technological advances led to widespread social change as we passed through the twentieth and entered the twenty-first century. Radio communications offered dramatic new opportunities for both personal communications and for commercial broadcasting. The harnessing of the radio spectrum for message carrying was a remarkable development in our social environment. Suddenly, anyone with the appropriate equipment could gain access to the home of anyone within their broadcast region, allowing them to inform, entertain, or to sell goods or services to the listener in the comfort of their own home. Commercial radio broadcasting began in the US in 1920 with the launch of KDKA Pittsburgh.[30] The UK quickly followed and in 1922 the BBC made its first broadcast using Marconi's 2LO transmitter at Marconi House in the Strand.[31] Early radio broadcasters could carry almost any content without state interference, but this did not mean they were unregulated. By 1926 there were 536 broadcasting stations in the US (in addition to over 17,000 amateur and maritime stations). These 536 stations had to share 89 wavelengths, leading to the prospect of jammed airwaves caused by interference between stations.[32] Competition between regulatory modalities was occurring. The market had supported the development of these myriad stations and without interference the market would doubtlessly have supported several hundred more radio stations.[33] The physical environment though would not allow the market to function in this way. Thus the design regulation inbuilt into the radio spectrum came into conflict with market imperatives. The impasse required intervention. Consequently, the US Congress passed the Radio Act 1927, creating the Federal Radio Commission and vesting in it control

30 Stark, P (1994) 'A History of Radio Broadcasting', *Billboard*, Nov 1 1994.
31 Media UK, *History of UK Radio* at *www.mediauk.com/article/8*.
32 Goodman, M, 'The Radio Act of 1927 as a Product of Progressivism', 1999, 2 *Media History Monographs www.scripps.ohiou.edu/mediahistory/mhmjour2–2.htm*.
33 On 13 April 2006, 100,000 watts, the US TV and Radio Directory (*www.100000watts.com/*) carried listings for 9,820 FM radio stations and 5,033 AM radio stations broadcasting within the 50 States.

over the radio spectrum. The FRC arose not as part of a considered policy to harness ownership (or property rights) as part of a hierarchical control network, but as a practical settlement of an ongoing regulatory conflict. As in many cases to follow, the federal government had settled an ongoing regulatory conflict by imposing hierarchical control. Once the required regulatory settlement is achieved though, the new regulator seeks to establish justification for their continued existence. Thus the radio industry became a closely regulated industry with the FRC (later the FCC) exerting control over politically sensitive content[34] and profanity,[35] leading eventually to the promulgation in 1949 of the fairness doctrine, which attempted to establish a standard of 'fair' coverage of important public issues.[36]

A similar model may be seen in the development of cinema regulation in the US where a self-regulatory approach was enshrined in the Motion Picture Production Code or 'Hays Code'. The Code was developed almost contemporaneously with the Radio Act. The regulatory conflict it was designed to settle was somewhat different. Whereas the Radio Act was designed to settle a competition/design conflict, the Code was designed to settle a community/competition conflict. In the 1920s the film industry was caught between the pursuit of box office success and the threat of censorship. Financial gains were to be made by producing racy and sexually suggestive films. In response to the development of such movies, religious groups called for content controls, even censorship. In 1922 the film industry attempted to self-regulate and put former Postmaster General Will Hays firmly at the centre of the fray. The self-regulatory Hays' Office attempted to head off censorship by cajoling producers to tone down content and by soliciting support for their actions from those groups demanding a more thorough crackdown. This attempt to clean up the industry was though far from successful in the eyes of community groups and state censors. On screen, actresses such as Clara Bow and Greta Garbo continued to portray storylines that challenged compliance with the Hays' Office,[37] while off-screen the actions of stars such as Wallace Reid, Mary Pickford, Fatty Arbuckle and Douglas Fairbanks were deemed

---

34 In 1928 the FRC renewed the licence for WEVD, owned by the Socialist Party, only with the stern warning that the New York station must 'operate with due regard for the opinions of others'. Federal Radio Commission, Order, 2 FRC 156 (1928).

35 Stark, see above, fn 30.

36 The fairness doctrine, which was applied to both radio and TV stations was eventually abandoned in 1987. See Hazlett, T and Sosa, D, 'Chilling the Internet? Lessons from FCC Regulation of Radio Broadcasting', 1997, *Cato Policy Analysis No. 270. www.cato.org/pubs/ pas/pa–270.html*.

37 Bow's performances in films such as *Hula* (1927) and *Red Hair* (1928) were felt to be morally questionable as was Garbo's performances in *Flesh and the Devil* (1927) and *Mysterious Lady* (1928).

inappropriate.[38] In 1930 the conflict between the commercial demands of the industry (which had been severely affected by the 1929 stock market crash) and the community values of its critics reached a head when Iowa Senator Smith Brookhart introduced a Bill designed to put the movie industry under the jurisdiction of the Federal Trade Commission.[39] The pressure forced action. In discussion with the industry's critics Will Hays helped draft the Motion Picture Production Code. By March it was adopted by the Association of Motion Picture Producers and the Motion Picture Producers and Distributors of America. With the Code in place the threat of Senator Brookhart was headed off, but at a cost.[40]

Like the introduction of the Radio Act, the Motion Picture Production Code did not arise as part of a considered policy to develop and improve the product on offer in cinemas. It arose in response to a particular regulatory conflict. While the market interests of the industry were in favour of greater freedom of content, the community standards represented by religious groups, state representatives and other pressure groups were in favour of greater external controls. The Code did not rise out of a regulatory vacuum, but from extended regulatory conflict. Like the Radio Act it was an imposed settlement that used hierarchical controls (through the Production Code Administration Offices managed by Joe Breen) and community values to develop a hybrid regulatory model. Both these historical

38 Wallace Reid, star of movies such as *Double Speed* and *Excuse My Dust*, became addicted to morphine in the early 1920s and died in a sanatorium on 18 January 1923. Mary Pickford was probably the first movie superstar. By 1916, aged 24, she was generally acknowledged as the most famous woman in the world. Mary married actor Owen Moore in 1911, but in 1917 she embarked on an affair with (also married) actor Douglas Fairbanks. In 1920 they divorced their partners, to marry. Then in 1922 she was romantically linked to director William Desmond Taylor, who was murdered in his Hollywood home. The worst scandal was certainly the Fatty Arbuckle scandal. Arbuckle was the star of the 'Fatty and Mabel' series of films. In 1921 he was charged with the manslaughter of a 25-year-old starlet named Virginia Rappe at a drunken party on Labor Day weekend. After three trials (the first two resulted in hung juries) Arbuckle was acquitted. It now appears that Miss Rappe died of peritonitis and Arbuckle had been the victim of a terrible smear campaign. At the time though, the scandal was incredible.
39 The 'Brookhart Bill' sought 'to eliminate the industry practice of blockbooking and blind-selling'. For discussion of the Bill see Mjagkij, N, 'The Film Industry in the United States: A Brief Historical Overview', *www.uwgb.edu/teachingushistory/2003_lectures/mjagkij_essay_1.pdf*, p 12.
40 The Code was famously restrictive. Among things banned were 'obscenity in word, gesture, reference, song, joke, or by suggestion', 'Pointed profanity including the words, God, Lord, Jesus, Christ – unless used reverently – hell, S.O.B., damn, Gawd, and every other profane or vulgar expression however used' and 'complete nudity, including nudity in fact or in silhouette, or any lecherous or licentious notice thereof by other characters in the picture'. For a general discussion of the Hays Code and the background to its adoption see Nizer, L, *New Courts of Industry: Self-Regulation Under the Motion Picture Code*, 1935, New York: Longacre Press.

developments are worthy of consideration today. The internet, like radio communications and cinema, is a socially disruptive technology. The competition/design conflict, faced by Congress in the 1920s with regard to radio is now being played out in cyberspace with regard to spam messages.[41] Spammers, like early commercial radio stations, compete for the attentions of customers. Unfortunately though, as with early radio, they are stretching available resources to the limit.[42] The community continues to make calls for effective regulation, but unlike the Radio Act, attempts by government to introduce hierarchical controls are failing. The EU Directive on Privacy and Electronic Communications has had limited impact on the number of spam messages being sent and received in Europe,[43] while in the US the promulgation of the Can Spam Act led to an increase in the levels of spam communications being sent.[44] Similarly, a twenty-first century version of the community/competition conflict, which threatened the 1920s film industry, is being played out in cyberspace. If the profitability of sexually suggestive material was clear to the Hollywood producers of the 1920s, by the new millennium sexual suggestiveness had given way to a mainstreaming of adult content. The cyberporn industry turns over more than $1 billion per annum.[45] Analysis carried out by Neilsen/Netratings estimated that in August 2003 31 million Americans – about one in four internet users in the US – visited adult websites,[46] while web-filtering company N2H2, estimated in September 2003 that there were around 1.3 million sites offering about 260 million pages of erotic content on the web.[47] Understandably, while

41 Attempts to regulate spam include the US Federal Can Spam Act 2003 (s 877); the EU Directive on Privacy and Electronic Communications Dir. 2002/58/EC; and the Personal Information Protection and Electronic Documents Act 2000 of Canada.

42 Symantec, the leading anti-spam service recorded that in the first six months of 2006, 61% of all emails sent were spam messages. At 2005 estimates this equates to around 20 billion spam messages per day.

43 On 22 January 2004 the European Commission noted that: 'While adopting legislation is a first, necessary step, legislation is only part of the answer. There is no "silver bullet" for addressing spam.' *Communication from the Commission on Unsolicited Commercial Communications or 'Spam'* COM(2004) 28 final, p 3.

44 A Brightmail press release on 2 February 2004 recorded: 'that 60% of all Internet email sent in January 2004 was spam. That represents a 2% increase from the company's measurement of 58% in December 2003.' See *www.brightmail.com/pressreleases/020204_can-spam-impact.html*. By April 2004 this figure had increased to 64% and had by 2006 settled at around 61%.

45 Thornburgh, D and Lin, H, *Youth, Pornography, and the Internet*, 2002, Washington DC: National Academies Press, p 72.

46 CNN, *Sex Sells, Especially to Web Surfers*, 11 December 2003. Available at: *www.cnn.com/2003/TECH/internet/12/10/porn.business/*.

47 *N2H2 Reports Number of Pornographic Web Pages Now Tops 260 Million and Growing at an Unprecedented Rate*. Available at: *http://ir.thomsonfn.com/InvestorRelations/PubNewsStory.aspx?partner=6269&layout=ir_newsStoryPrintFriendly.xsl&storyid=94774*.

the adult services industry profits from one of the few successful internet business models, community groups have reacted strongly. Groups such as Citizen Link,[48] the American Family Association[49] and Pure Intimacy[50] have called for intervention. Governments have acted in an attempt to impose a solution to these conflicts. The US Communications Decency Act of 1996 was the first, failed, attempt to do so.[51] This was followed in the US by the Child Online Protection Act 1998 and the Children's Internet Protection Act 2000, both of which have been of limited success.[52] Similar attempts to regulate content in Europe and Australasia have been equally ineffectual at reducing the quantity and explicitness of pornographic material accessible via the web.[53] These extant regulatory conflicts are proving much more difficult to settle than the radio or motion picture conflicts of the 1920s. The reasons for this are numerous, but two vital distinctions stand out. The first of these is the global design of cyberspace. As illustrated by the cyber-libertarian school, individuals and corporations may evade state-based control in cyberspace by utilising the network architecture.[54] This severely impairs the ability of a state-based regulatory body to impose a settlement: the traditional regulatory model of imposing a hierarchical settlement appears to be ineffective in cyberspace. This failure means that unless alternative regulatory settlements can be found, such conflicts may theoretically rage until the consumption (or misappropriation) of resources destroys the utility of the network. The second distinction both provides the alternative regulatory model sought, and introduces a new set of challenges for the cyber-regulator. As outlined by Lawrence Lessig, in cyberspace we have a unique level of environmental control through manipulation of its code.[55] Thus where traditional state-based regulatory settlements fail due to the trans-jurisdictional nature of the space, the very fact that we control the

---

48 The Citizen Link website is *www.family.org*.   49 The AFA website is *www.afa.net/*.

50 The Pure Intimacy website is *www.pureintimacy.org/*.

51 The Communications Decency Act was struck out by the Supreme Court in *Reno v American Civil Liberties Union* 117 S Ct 2329 (1997). It is discussed in depth in ch 6.

52 The Child Online Protection Act was challenged and suspended in *Ashcroft v American Civil Liberties Union* 535 US 564 (2002). This decision was confirmed by the US Supreme Court on 29 June 2004 where by a majority of 5–4 the Court found the Act to be unduly restrictive of the first amendment right to Free Speech. See *Ashcroft v American Civil Liberties Union* (03–218) 29 June 2004. The Children's Internet Protection Act successfully survived a first amendment challenge in Summer 2003 in the case of *United States v American Library Association* (02–361) 23 June 2003.

53 Eg Resolution on the Commission communication on illegal and harmful content on the Internet (COM(96)0487 – C4–0592/96) (EU); The Broadcasting Services Amendment (Online Services) Act 1999 (Aust).

54 Johnson, D and Post, D, 'Law and Borders – The Rise of Law in Cyberspace', 1996, 48 Stan LR 1367.

55 Lessig, see above, fn 23, pp 100–08.

environment of that space in a hitherto unparalleled manner, allows for a different form of regulatory settlement – one based not in a hierarchically imposed settlement, but rather one based in design. The key to understanding cyber-regulation is therefore in understanding its environmental plasticity.

Socio-legal regulation and socio-technical-legal regulation

Traditional regulatory discourse has been informed by an understanding of law and society. The study of the interaction between legal regulation and social change has been central to much western European legal research in the last 50 years. Led initially by eminent US scholars such as Philip Selznick, Lon Fuller and Robert Kagan, European scholars such as David Nelkin, Michael Zander, Denis Galligan, David Sugerman and Andre-Jean Arnaud were soon following in the footsteps of early European law and society scholars such as Hermann Mannheim. The study of law and society (or socio-legal studies) recognises that law is a social science and as such cannot be examined in a vacuum. Thus the actions of lawmakers, lawyers, legal scholars and commentators form part of a social matrix. Regulators form part of this matrix and, as we have traditionally modelled regulatory designs upon the belief that active choices are made by governments or other regulatory bodies, we have used this matrix as a key component of regulatory design and regulatory theory. A topology of traditional socio-legal approaches to regulation was given by Mark Thatcher.[56] Thatcher suggested that the traditional approaches to regulatory design may be categorised into five models.

The first model is *Public Interest*. This is the oldest approach to regulation, involving a functionalist perspective with an apolitical ontology. That is, state action is considered a means to correct market failures, or in the 'public interest', managing natural monopolies, public goods, and externalities that may arise (such as in the area of defence, or pollution issues), informational problems and transaction costs. This is a paternalistic approach where government is assumed to have more information than other actors. Regulation is considered a positive development in its own right. There is often little explanation, however, of how the 'public interest' is actually achieved through regulation. This is symptomatic of a lack of regard of the self-interest of government and the regulators, similar to the problems with the technology and economics literature. Second, there is the *Capture/Cyclical* model. This model arose as political scientists noted that in some cases the regulators and the regulated achieved a level of consensus. This occurred particularly when regulatory agencies were captured by the interests

56 Thatcher, see above, fn 22.

of regulatees. In this view, regulations are a means of protecting the interests of the regulated industries rather than as a means of protecting the interests of consumers and new entrants into a market.[57] This theory is weak, however, at explaining why some regulations fail to meet the interests of the regulated industries, as it lumps together the interests of the regulator and the regulatees uncritically. Even within a given industry affected by a regulation, a number of interests may exist; this theory presupposes monolithic interests that may be embedded within a regulator, something that Christopher Hood qualifies as 'simplistic'.[58] It also pays insufficient attention to the interests and actions of other actors such as the courts, consumers and voters. Capture theory is, though, an improvement from the public interest model as it considers the politics of regulating, and translations of interests. Thatcher's third model is *The Economic Theory*. The economic theory of regulation is associated with the Chicago School.[59] Regulation is seen as arising from the interaction between rent-seeking interest groups demanding regulation and politicians supplying it. It is argued that all forms of regulation actually disadvantage consumers, as interest groups pursue regulatory rents, and these groups tend to be producers or powerful users. This approach allows for an analysis of specific interests in specific actors. The school's view of interests, however, is criticised as being too narrow as it focuses only on material self-interest, failing to see other possible interests.[60] Fourth, there is *Public Choice Theory*. Public choice theory separates politicians from regulators, seeing even more actors make up the discourse. Politicians are seen as a group of actors interested in maximising budgets, distributing patronage, or perhaps instead as actors seeking smaller government through tax cuts. Bureaucrats and regulators are seen as actors pursuing larger budgets, or wishing for autonomy, power and prestige. Approaching regulation studies in this way permits an elicitation of numerous motivations behind a given regulatory regime, beyond simpler rents like money and votes. This view does not, however, explain deregulation adequately: if interests are aligned behind a regulatory regime, there is little reason for regulatory change.[61] Finally there is the model of *Historic Institutionalism*. In this theory, institutions are the most important units of analysis. That is, institutions are stable and have a history, and in turn affect which actors have power. This view helps to explain why different states develop different

---

57 See also Baldwin, R and Cave, M, *Understanding Regulation: Theory Strategy and Practice*, 1999, Oxford: OUP, p 36.
58 Hood, see above, fn 2, p 22.
59 For a discussion of the Chicago School see the discussion in Chapter 1 and G. Stigler, *Chicago Studies in Political Economy*, 1988, Chicago: Chicago UP.
60 Hood, see above, fn 2, p 24.
61 Noll, R, 'The Economic Theory of Regulation after a Decade of Deregulation: Discussion piece' in Baldwin, Scott and Hood, see above, fn 2, p 140.

interests and in turn different regulations. In a move beyond economic and public choice theories, interests are seen as emerging from a context, rather than something to be assumed. While this theory can explain differences among regimes, it suffers from failing to explain regulatory trends and changes. These models of regulatory design, like the models of regulatory methods or modalities discussed earlier, assume positive intervention based upon a policy strategy and as a result reflect the traditional regulatory approach of seeking a socially mediated settlement to such a disruption. As any alteration to the regulatory settlement occurs through a change in the socio-legal modalities of regulation this *appears to be* a policy decision, when in fact this is not always the case. As we have seen though, changes to the regulatory settlement are often not the result of a change in policy, but rather are formulated by the need to solve a regulatory disturbance caused by a disruption to the settled position. In understanding where the drive for change originates, we can develop a more precise model to allow us to understand regulatory design.

To understand why regulatory design is different in the sphere of the new media we have to differentiate between socio-legal modalities of regulation, as applied in the physical world, and socio-technical-legal (STL) modalities as employed in informational space. If we return to Lawrence Lessig's modalities of regulation thesis we see that Professor Lessig suggests the application of four modalities: Law, Market, Norms and Architecture. These modalities may in turn be categorised into two families: (1) socially-mediated modalities and (2) environmental modalities (see figure 2.1).

In physical space environmental modalities suffer from a high degree of inertia: inertia integral to the physical laws of the universe. This inertia can be most clearly illustrated by the second law of thermodynamics, which states that a closed system will remain the same or become more disordered over

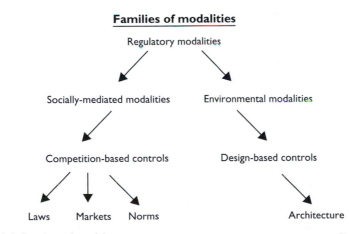

Figure 2.1 Families of modalities.

time: in other words its entropy will always increase.[62] The natural order of
the universe is that our physical surroundings (environment) become less
regulated over time. To overcome this, or in other words to harness design-
based controls, we must bring an external force to bear. Thus environmental modalities are resource-intensive: to utilise an environmental modality
the regulator must expend considerable initial resources to overcome this
universal inertia (see Figure 2.2).

For this reason in those areas of regulatory policy where environmental
modalities have traditionally been used, such as transport policy, we see a
large proportion of the regulator's resources being expended on design and
construction. The development and construction of controls such as road
humps, one-way systems, directive road layouts and traffic control systems
consume a considerable part of the transport planners annual budget.[63] Even
simple design-based systems such as the Inland Revenue's self-assessment tax
regime consume a considerable amount of resources in their implementation,
although it should be recognised that the employment of design-based
systems is often self-financing in the longer term, due to the self-enforcing
nature of most such controls.[64] Despite this, the large initial investment
required to *overcome the environment* often mitigates against the extensive
use of environmental modalities in the regulation of physical space. The
employment of socially mediated modalities, by comparison, does not usually
require the overcoming of such inertia. Thus in designing regulatory

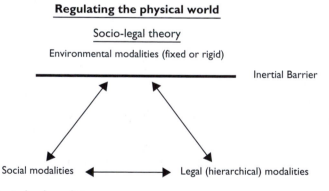

Figure 2.2 Socio-legal regulation.

62 For discussion of the Second Law of Thermodynamics see Kuhn, T, *Black-Body Theory and
the Quantum Discontinuity: 1894–1912*, 1978, Chicago: University of Chicago Press.
63 Figures supplied by Hertfordshire County Council suggest that road humps cost on average
£1,500 per installation, while road narrowing costs approximately £5,000 per narrowing, the
installation of a Pelican crossing costs on average £18,000 and the creation of a 20mph zone
can cost as much as £100,000.
64 Enforcement systems will be discussed below. Environmental modalities that harness the
physical laws of nature are self-enforcing and therefore more cost-effective once the initial
implementation costs have been met.

structures for the physical world we usually give pre-eminence to socio-legal (or socially-mediated) modalities of regulation.

We can see examples of the application of socially mediated modalities in many everyday regulatory settlements. For example, we can see this if we examine the early history of broadcast regulation in the UK. Broadcast technology first arrived in the UK in February 1896 when Guglielmo Marconi arrived in London with his wireless transmitter technology.[65] Marconi developed his technology at the Hall Street Works, Chelmsford, which were acquired by the Wireless Telegraph and Signal Company in 1897 as the site of the world's first factory for the production of radio transmitters and receivers. In 1900 Marconi made a further breakthrough: the famous '7777' patent granted on 26 April 1900. This documented a system that allowed simultaneous transmissions on different frequencies. Adjacent stations were now able to operate without interfering with one another and ranges were increased. With the success of Marconi the first regulatory conflict arose. The potential public value of the broadcast spectrum had been made clear in 1899 when a wireless message was received from the East Goodwin lightship. It had been rammed in dense fog by the steamship *R.F. Matthews* and made a request for the assistance of a lifeboat. A lifeboat was swiftly launched to go to the assistance of the lightship, though fortunately the damage was not serious and she remained on station. This was the first successful maritime distress call by wireless. With an ever-increasing demand being placed upon the available broadcast spectrum by radio enthusiasts though, there were concerns as to whether the public interest would be adequately protected. The available broadcast spectrum was limited, whereas demands upon it were increasing, apparently without limit. The government was faced with a conflict between community values, competition and the environmental limits of the available spectrum. The response was to implement a hierarchical/competition solution. Through the Wireless Telegraphy Act 1904, all UK wireless telegraphy users were required to obtain a licence for transmitting and receiving. Further, the Act regulated all wireless communication to ensure that the entire radio frequency spectrum would be used for the public good, and to allow development and enforcement of international agreements. This Act created a new regulatory regime. It imposed a hierarchical settlement and created a new policing and enforcement role for the Post Office. In addition, as licences required to be paid for, it introduced a new market condition into the nascent wireless market, thus harnessing the modality of competition to provide an additional layer of regulation. An alternative, and arguably more efficient, regulatory settlement would have been to harness the environment and formulate a design-based solution. Many design solutions were theoretically possible: one could for example

---

65 It may be noted that Marconi received a patent on his technology (patent number 12039), which was filed on 2 June 1896.

increase the available spectrum through harnessing alternative bandwidth, or one could increase the efficiency of use of the available spectrum through technologies such as spread spectrum, digitisation or packet switching. But, due to the high degree of environmental inertia that these solutions were required to overcome, such design-based solutions were not possible in 1904. With the later development of technologies that allowed us to overcome this inertia, such as FM transmission,[66] and the development of new economic models for commercial radio stations, the UK Government could abolish the licence (and fee) for radio transmission and reception in 1971, demonstrating that in the longer term, design-based solutions may provide the most efficient method of regulation. In the real world though, they cannot match the plasticity of socially mediated solutions in the short-to-medium term.

Having established that there is a natural pre-eminence for socially mediated regulatory modalities in the physical world, we can model how they function and why they function in a more flexible manner than design-based solutions. As socially mediated modalities are competition-based modalities, and further as regulatory settlements are usually the result of competition between regulators or regulatory modalities, the regulatory authority called upon to settle such regulatory disruption may choose to utilise any of the socially mediated modalities either alone or in a hybrid regulatory model and they may do so in one of two ways: either to allow one party to claim victory, or to declare the other party disqualified.[67] Although this may on the surface seem the same thing, the effects are in practice quite different. The distinction is seen most clearly from the point of view of a neutral third party. Whereas the effect of the settlement on the parties in dispute may be the same, the outcome of the settlement may look quite different to our third-party observer. We can see this if we imagine the situation, wherein the issue in conflict is the availability and distribution of indecent images on the internet. The conflict here is between the market imperatives of the adult services industry and the community imperatives represented by groups such as Citizen Link.

In a regulatory competition with only two (opposing) views, there are four possible outcomes: (1) Allow or legitimise indecent content, (2) Disqualify or outlaw indecent content, (3) Defend or standardise Citizen Link community standards, or (4) Disqualify or ignore Citizen Link community standards. At first sight outcome one and outcome four seem to achieve the same result: the allowing of indecent content, while outcomes two and three seem to share the same result of removing or banning such content. In fact this is not the case. Outcome one is an example of positive, or permissive, regulation. Here we allow the adult services industry to continue to trade, should they so wish. Outcome four meanwhile is a negative, or preventive, regulation restricting

---

66 'Yankee Frequency Modulation About Ready: Armstrong Method To Go on Air in June', *Broadcasting*, 1 June 1939.

67 The option of a 'tie' or 'draw' is discounted as this would perpetuate the dispute.

the freedom of a particular social grouping. To our third-party observer the distinction becomes clear. If the regulator chooses to impose a positive settlement based upon outcome one, our third party has no opportunity as an external observer to then bring about a later challenge against the suppliers of such content as the content has been permitted. If, though, the regulator chooses to implement the fourth outcome and disqualify the challenge of the Citizen Link community, our third-party observer may bring about an independent challenge, based upon different community values or standards. Thus while the first outcome disempowers our third-party observer, outcome four does not: this is because the first outcome is positive (or green light) regulation while the fourth is negative (or red light).[68] The flexibility offered by combining permissive and preventive regulation allows regulators, when dealing with competition-based regulatory disputes, to design regulatory settlements that will be least disruptive and therefore most open to acceptance from all parties to the dispute. In this way the inertia that exists in socially mediated modalities may be overcome. You cannot, though, negotiate with the universe (design-based solutions always face the inertia predicted by the second law of thermodynamics), which is why socio-legal settlements have pre-eminence in the physical world. Cyberspace though is an entirely virtual environment; it has no physical presence beyond its wires and routers. Once one ventures into the content layer of cyberspace the environmental inertia obligated by the second law of thermodynamics no longer applies.[69] This release allows for a new flexibility in the relationship between law, society and design. It is this that forms the basis of socio-technological-legal theory (STL). With the inertia of the physical laws overcome we can map a new regulatory model in which environmental modalities are equally functional with social-mediated modalities (see Figure 2.3). In the STL model we can exploit regulatory settlements that *design the environment*.

Understanding that regulatory discourse may include technology is another step in understanding regulation. The need to consider technology in regulation is not new. While it may be acknowledged as a disruptive force, as Peltzman notes within telecommunications rate structures and interest rate regulation,[70] it traditionally has not been investigated in detail. Levin notes with regard to radio spectrum management, that 'new communications have required the re-examination of many policies and assumptions in recent years',[71] while Porter and Van der Claas found that the literature on

68 For a discussion of green-light and red-light theories in administrative law see Harlow, C and Rawlings, R, *Law and Administration*, 2nd edn, 1997, London: Butterworths.
69 Network Layers will be discussed in more depth below.
70 Peltzman, S, 'The Economic Theory of Regulation after a Decade of Deregulation' in Baldwin, Scott and Hood, see above, fn 2.
71 Levin, H, 'New Technology and the Old Regulation in Radio Spectrum Management', 1966, 56(1/2) *American Economic Review* 339.

Regulating technology-based platforms

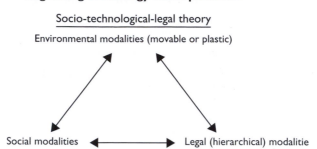

Figure 2.3 Socio-technical-legal (STL) regulation.

environmental regulation tended to assume a static view of the technology, products, and processes.[72] This static approach was apparent in the early literature on cyber-regulation. The cyber-libertarian approach was predicated on the concept that regulatory authorities would be unable to enforce their regulatory standards within the network due to the inherent design features of the internet, which would render any attempts at regulatory intervention futile. With the development of the cyber-paternalist school the malleable nature of the network architecture became a central pillar of cyber-regulatory theory. Lawrence Lessig in particular seized upon the true nature of the network architecture in his examination of cyber-regulatory modalities in *Code and Other Laws of Cyberspace*.[73] Here Lessig's famous maxim that *Code is Law* is fleshed out with Lessig proposing greater inclusion of technology into regulatory discourse. Lessig argues that architecture can and does regulate individuals. Building from Bentham and Foucault's conceptions of the Panopticon, he shows how structures regulate.[74] In particular he demonstrates the fluidity of the environment in Cyberspace. He names the default internet structure of the time Net95, and demonstrates that although the current structure of the network may allow or disallow certain transactions, such as identification of individuals, it does not mean that this default position is always enshrined within the network protocols.[75] As Lessig demonstrates, there were times these other constraints were treated as fixed: when the constraints of norms were said to be immovable by governmental action, or the market was thought to be essentially unregulable, or the cost of changing architecture was so high as to make the thought of using it for

---

72 Porter, M and van der Claas, L, 'Towards a New Conception of the Environment-Competitiveness Relationship', 1995, 9(4) *Journal of Economic Perspectives* 97.
73 Lessig, see above, fn 23, pp 85–99.   74 Ibid, pp 91–3.
75 For example Lessig demonstrates the power of digital certificates to identify and regulate individuals. Ibid, pp 49–53.

regulation absurd. But we see now that these constraints are plastic. That they are, as law is, changeable, and subject to regulation.[76]

Environmental layers and network architecture

Having recognised that there is a particular role for technology within design-centred cyber-regulatory discourse in the post cyber-paternalist model, we must examine the practical application of technology as an essential element of the STL regulatory settlement for cyberspace. The key to this is in examining the environmental layers of the network architecture. Stratification, or layering, may be identified in any informational environment. Both network engineers and communications theorists recognise the vital function played by environmental layers in communications networks. In his book *Weaving the Web*,[77] the architect of the World Wide Web, Tim Berners-Lee, identifies four layers within the architecture of the web: the transmission layer, the computer layer, the software layer and the content layer.[78] This may be seen as a simplified version of the seven-layer open systems interconnection reference model (OSI model) used in all network design. This model divides the functions of a protocol into a series of layers. Each layer has the property that it only uses the functions of the layer below, and only exports functionality to the layer above. A system that implements protocol behaviour consisting of a series of these layers is known as a 'protocol stack'. Protocol stacks can be implemented either in hardware or software, or a mixture of both. Typically, only the lower layers are implemented in hardware, with the higher layers being implemented in software.[79] Although both the OSI model and the

---

76 Lessig, above, fn 75, p 91.
77 Berners-Lee, T, *Weaving the Web: The Original Design and Ultimate Destiny of the World Wide Web by Its Inventor*, 2000, San Francisco: HarperCollins.
78 Ibid, pp 129–30.
79 The seven OSI layers are: (1) Physical layer: the major functions and services performed by the physical layer are: establishment and termination of a connection to a communications medium and participation in the process whereby the communication resources are effectively shared among multiple users. (2) Data link layer: the data link layer provides the functional and procedural means to transfer data between network entities and to detect and possibly correct errors that may occur in the physical layer. (3) Network layer: the network layer provides the functional and procedural means of transferring data sequences from a source to a destination via one or more networks while maintaining the quality of service requested by the transport layer. (4) Transport layer: the transport layer provides transparent transfer of data between end-users. (5) Session layer: the session layer provides the mechanism for managing the dialogue between end-user application processes. It establishes checkpointing, adjournment, termination and restart procedures. (6) Presentation layer: the presentation layer relieves the application layer of concern regarding syntactical differences in data representation within the end-user systems. (7) Application layer, the highest layer: this layer interfaces directly to, and performs common application services for, the application processes. See Comer, D, *Internetworking with TCP/IP: Principles, Protocols and Architecture*, 2000, Paramus, NJ: Prentice Hall.

Berners-Lee model are network architecture models, designed to describe the purely functional aspects of the network, they can easily be adapted to illustrate the challenges faced by regulators. For regulatory theorists this has been most successfully done by Yochai Benkler in his eloquent paper *From Consumers to Users*[80] and it is that model to which we must turn.

Benkler describes a three-layer network that is similar to Berners-Lee's four-layer environment. Benkler labels his layers: (1) the physical infrastructure layer, (2) the logical infrastructure layer, and (3) the content layer. What Benkler does is to reduce the OSI/Berners-Lee model to the three-key regulatory access point within the network. He does so by combining Berners-Lee's transmission layer and computer layer into a single physical infrastructure layer. Benkler's three layers represent the three vital environmental layers found on the internet. The foundational layer is the physical infrastructure layer. This is the only network layer that is subject to the environmental regulations of the physical world, such as the second law of thermodynamics. The physical infrastructure layer is the link between the physical world and cyberspace and is made up of wires, cables, spectrum and hardware such as computers and routers.[81] The second layer, the logical infrastructure layer, encompasses the necessary software components to carry, store and deliver content, software such as the TCP/IP protocol, SMTP, HTTP, operating systems and browsers. Finally, the content layer encompasses all materials stored, transmitted and accessed using the software tools of the logical infrastructure layer. Benkler's model was adapted by Lawrence Lessig in his book *The Future of Ideas*. Lessig rebranded the layers the physical layer, the code layer and the content layer; this allowed him to discuss the particular effectiveness in using the code layer to regulate the content layer, a subject he had previously raised in *Code and Other Laws of Cyberspace*, and to which he would return in his third book, *Free Culture*.

Both Benkler and Lessig highlight the effectiveness of introducing regulation into the network through the logical infrastructure or code layer. Benkler noted that the Digital Millennium Copyright Act (DMCA) 'permits the owners of copyright to design the logical layer of the distribution media of their work to assure that their works are perfectly protected by technology, irrespective of whether the uses that users are seeking to make of these works are privileged by law'.[82] Lessig similarly notes that, 'using code copyright owners restrict fair use; using the DMCA, they punish those who would attempt to evade the restrictions on fair use that they impose through code.

---

80 Benkler, Y, 'From Consumers to Users: Shifting the Deeper Structures of Regulation Toward Sustainable Commons and User Access', 2000, 52 Fed Comm LJ 561.
81 A router is a special-purpose computer or software package that handles the connection between two or more networks. Routers spend all their time looking at the destination addresses of the packets passing through them and deciding which route to send them on.
82 Benkler, see above, fn 80, p 571.

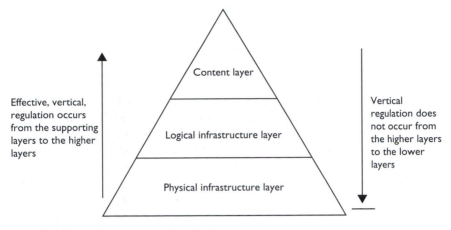

Figure 2.4 Vertical regulation and Benkler's layers.

Technology becomes a means by which fair use can be erased; the law of the DMCA backs up that erasing. This is how *code* becomes *law*.'[83] The key to Benkler's model is through the recognition that by introducing a regulatory modality at one of these layers you may vertically regulate, but that such vertical regulation is only effective from the bottom-up, that is regulation in a supporting layer is effective in the layers above, but does not affect the layers below (see Figure 2.4). This is because the higher layers rely upon the infrastructure of the lower layers to be effective. The high-level content found in the content layer is facilitated by the lower-level protocols of the logical infrastructure layer, which in turn is reliant upon the effectiveness of the physical infrastructure layer. By amending the lower layers, the upper layers may effectively be regulated. Without reciprocal reliance though the reverse is not true, an amendment in the content layer has no direct effect on the logical infrastructure layer. This was first recognised by Lessig in his book *The Future of Ideas.* Lessig realised that the internet protocols (the code layer) only developed because the telephone companies (the controllers of the physical layer) did not intervene in the development of the network. As Lessig notes, 'what is striking about the birth of this different mode of communication is how little the telephone companies did in response. As their wires were being used for this new and different purpose, they did not balk. They instead stood by as the internet was served across their wires'.[84] This came about, as noted by Lessig, due to the intervention of the Federal

83 Lessig, L, *Free Culture: How Big Media Uses Technology and the Law to Lock Down Culture and Control Creativity*, 2004, New York: Penguin, p 160 (original emphasis).
84 Lessig, L, *The Future of Ideas: The Fate of the Commons in a Connected World*, 2001, New York: Random House, p 148.

Communications Commission (FCC) who prevented the telephone companies from interfering with the development of the new technology.[85] Further, he noted that a similar effect could be recorded at the code/content interface. In discussing the development of the web, Lessig notes that 'other languages (than HTML) enabled authors to easily hide the code that makes the scripts run. But had it been the default, the knowledge commons of the World Wide Web would have been vastly smaller . . . free code ensures that innovation cannot be chilled . . . free code hasn't the power to discriminate against new innovation.'[86] Thus Lessig clearly identifies the ability of the supporting layers to regulate the content of the higher layers. Lessig uses this to argue for greater freedom and openness in the supporting layers, making a strong case for a neutral platform or commons. This, according to Lessig, is the only way to protect cultural property creativity in the highest layer, the content layer.[87] Professor Lessig's argument in *The Future of Ideas* is erudite, but idealistic. It is highly unlikely that content producers, media corporations and other copyright holders will allow for a neutral system designed to protect cultural property and creativity at the cost of loss of control over their products. Thus any attempt to create a commons in the physical infrastructure and logical infrastructure layers does not create, or provide for, any regulatory settlement. In fact the introduction of any such system would merely create further regulatory competition, or conflict – the conflict which would arise in this situation being a competition/design conflict between the market imperatives of the content owners and the design of the environment within the supporting layers.[88] This conflict would though be further complicated by the use of hierarchical intervention to create the design environment. Thus in effect, what looks like a competition/design conflict would in fact be a competition/hierarchy/design conflict. The complexity of this conflict would almost certainly overshadow any benefits achieved by the imposed hierarchy/design settlement proposed by Professor Lessig. This is a salutary warning for any cyber-regulatory theorist who assumes that by imposing an environmental change a settlement may be achieved. In fact the harnessing of one regulatory modality through the application of another is more likely to lead to further regulatory competition, due to the complexity of the network environment.

Modelling regulation in cyberspace: Polycentric webs and layers

We can now begin the process of mapping cyber-regulatory modalities, their use and their potential effects. We have already identified two significant

---

85 See the discussion of this in Huber, P, Kellog, M and Thorne, J, *Federal Telecommunication Law*, 2nd edn, 1999, Frederick, MD: Aspen Publishing.
86 Lessig, see above, fn 84, p 58.   87 Ibid, p 266.
88 On the economics of intellectual property see Landes, W and Posner, R, *The Economic Structure of Intellectual Property Law*, 2003, Cambridge, MA: Belknap Press.

distinctions in cyber-regulatory theory as opposed to regulation in the physical environment: (1) increased utility of environmental regulation and (2) the complex effect of layers within communications networks. To this we can now add a third complication, which applies equally in physical space and cyberspace – decentred or polycentred regulation.

As has been alluded to earlier in this chapter, regulation rarely, if ever, emanates from a single source. The existence of polycentric regulation is the source of regulatory conflict, both between competing regulators and within a single regulatory body. Here I define polycentric regulation as 'the enterprise of subjecting human conduct to the governance of external controls whether state or non-state, intended or unintended'. This definition draws upon Lon Fuller's definition of law as 'the enterprise of subjecting human conduct to the governance of rules'[89] and the European approach to regulation which defines regulation as 'all forms of social control, state and non-state, intended and unintended'.[90] Both Fuller's definition of law and the wide definition of regulation leave ample room for overlapping, even conflicting regulatory systems. As Fuller himself noted, 'A possible . . . objection to the view [of law] taken here is that it permits the existence of more than one legal system governing the same population. The answer is, of course, that such multiple systems do exist and have in history been more common than unitary systems.'[91] The second half of Fuller's defence is the key to understanding polycentric regulation. While polycentric law was seen as quite a radical departure from the traditional model of a state monopoly in legal regulation of its citizens, it has never been the case that states have been able to claim any kind of general regulatory monopoly. This is due to the complexity of the environment, and a thriving market in regulatory competition. All the examples or regulatory competition discussed in the first half of this chapter are examples of competing regulators, and thus may be classified as examples of polycentric regulation. We can illustrate polycentric regulation through the application of a further example.

The People's Republic of China (PRC) is one of the most centralised and monocratic regulatory jurisdictions of the current age. The Communist Party of China (CPC) which has governed the PRC since 1949 is infamous for implementing a policy of far-reaching media control and censorship.[92] The development of new communications technologies, and in particular the

---

89 Fuller, L, *The Morality of Law*, 1964, New Haven, CT: Yale UP, p 106.
90 See above, fn 2.   91 Fuller, see above, fn 89, p 123.
92 Within the PRC there is heavy government involvement in the media, with many of the largest media organisations (namely CCTV, the People's Daily and Xinhua) being agencies of the Chinese Government. Also all Chinese media outlets are banned from questioning the legitimacy of the Communist Party of China, and the government often uses laws against state secrets to censor press reports about social and political conditions.

internet has been of particular concern to the CPC due to its expansive and unfiltered content. As a result, the CPC have made several regulatory interventions in an attempt to regain control over content. Probably the most wide-reaching of these is the creation of a state-supported firewall,[93] which has become known colloquially as the 'Great Firewall of China'. The system is quite sophisticated. It employs a matrix of controls, using a combination of filtering technology and keyword analysis,[94] and functions by preventing access to particular IP addresses through a standard firewall and proxy servers at the internet gateways. The government does not appear to be systematically examining internet content; instead, they are using keywords in site content and meta-descriptions to add sites to the blacklist. Analysis carried out in 2002 by Jonathon Zittrain and Benjamin Edelman at the Berkman Center for Internet & Society demonstrated the effectiveness of the system.[95] Despite extensive investment by the CPC in a design solution to the problem, they are finding it increasingly difficult to exercise control over media content within the borders of the PRC. Competing regulators continually interrupt their attempts to impose control. In April 2003, the US Government's International Broadcasting Bureau announced that it had commissioned Bennett Haselton of Peacefire.org to create a user-friendly software program to circumvent the Chinese firewall. In justifying their actions, Ken Berman, program manager for internet anticensorship at the Bureau said, 'the news is highly censored. The Chinese Government jams all of our radio broadcasts and blocks access by their people to our Website. We want to allow people there to have the tools to be able to have a look at it.'[96] The completed software, named 'Circumventor' was launched on 16 April 2003 and along with commercial systems such as Dynamic Internet Technology's DynaWeb system,[97] has provided a simple and accessible circumvention tool. It is hard to gauge the effectiveness of these technologies given the secretive nature of the PRC, but in an interview with Bennett Haselton on 20 May 2003 he reported that 'we've gotten e-mails from people using it in China',[98] while data from an earlier Chinese Academy of Social Sciences survey suggests that

---

93 A firewall is a piece of hardware or software put on the network to prevent communications forbidden by the network policy.

94 For a detailed discussion of filtering techniques see Deibert, R and Villeneuve, N, 'Firewalls and Power: An Overview of Global State Censorship of the Internet' in Klang, M and Murray, A (eds), *Human Rights in the Digital Age*, 2005, London: Glasshouse Press.

95 Zittrain, J and Edelman, B, 'Empirical Analysis of Internet Filtering in China', 2003, available at: *http://Cyber.law.harvard.edu/filtering/china/*.

96 Berman, K, interview quoted in Festa, P 'Software Rams Great Firewall of China', *CNET News*, 16 April 2003, available at: *http://news.com.com/2100–1028–997101.html*.

97 It should be noted DynaWeb was also developed with the support of the US Government. See *www.dit-inc.us/about.htm*.

98 Festa, P, 'Cracking the Great Firewall of China' *CNET News*, 20 May 2003, available at: *http://news.com.com/2008–1082_3–1007974.html*.

as many as 10 per cent of internet users in the PRC regularly use proxy servers to defeat censorship.[99]

What we see here are at least three competing regulatory bodies seeking to control access to information. The dominant regulator is the Government of the PRC, yet they do not hold a monopoly over controlling access to information in the PRC. They find themselves challenged by external government agencies such as the International Broadcasting Bureau and by private enterprises such as Dynamic Internet Technology and Peacefire.org. A simple map of the competing regulatory claims upon access to information in the PRC at this stage looks like this:

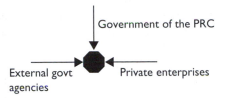

Figure 2.5 Simple regulatory model.

Here we can clearly see the simplest effects of polycentric regulation: no one regulator may impose their will on any subject of regulation without the agreement of competing regulators (and the support of regulatees). This model though only captures part of the regulatory conflict that currently surrounds regulation of access to content in the PRC. This only records regulators attempting to employ a hierarchy, design or hierarchy/design hybrid regulatory model. Other regulatory modalities have been used in the past by the CPC to control the availability of information in the PRC. Historically, the government attempted to control the market for information through ownership of major media outlets such as the *People's Daily* and CCTV, and by heavily subsiding all other media outlets to allow the CPC to exercise a degree of editorial control. Thus historically, the supply-side of the information market was controlled by the CPC. In recent years though this market has grown out of control and the CPC has been unable to keep control over the supply of information in the PRC. Most state media organisations no longer receive substantive government subsidies and are expected to largely pay for themselves through commercial advertising. As a result, they can no longer serve solely as mouthpieces for the government: they must produce programming that people find attractive and interesting. In addition, while the government does issue directives defining what can and cannot be published, it does not prevent, and in fact actively encourages, state media

---

99 The actual survey cannot be sourced in English, but it is referred to by Liang, G and Wei, B, 'Internet use and Social Life/Attitudes in Urban Mainland China', 2002, 1 *IT & Society* 238. The full title of the survey is *The CASS Internet Report (2000): Survey on Internet Usage and Impact in Five Chinese Cities.* Beijing: Chinese Academy of Social Sciences.

outlets to compete with each other for viewers and commercial advertising. Thus a diversity of informational products has arisen, making it more difficult for the CPC to exercise effective control. Further, technologies such as mobile telecommunications, and in particular SMS messaging, have created a new informational market, which the government is finding extremely difficult to control. With the weakening of central controls over media content the people of the PRC have become more sophisticated in their demands, with at least one report suggesting that the *People's Daily* saw its circulation reduced from 3.1 million copies in 1990 to 2.2 million copies in 1995, while local newspapers with a more diverse content saw a large increase in circulation in the same period.[100] Thus if we repeat the exercise of simply mapping the competing regulatory claims upon access to information in the PRC, from a purely market viewpoint, they look like this:

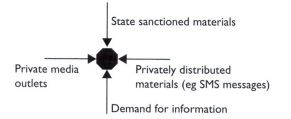

Figure 2.6 Simple regulatory model – market.

If we overlay Figures 2.5 and 2.6 we start to get an impression of the competing regulatory demands being made on the market for information in China:

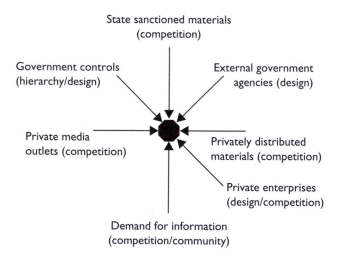

Figure 2.7 Information in China: Simple regulatory model.

100 *www.wordiq.com/definition/Media_in_China.*

Where several regulators vie for regulatory acceptance they do not act in a regulatory vacuum, any action by one has an effect on the actions of the others. This is because all regulators form part of an environmental system and a change in any one aspect of this environment affects all who participate in that environment. Thus it is wrong to imagine the regulatory subject, or 'pathetic dot'[101] as being a merely passive receiver sitting at the middle of a torrent of regulatory demands. Rather the regulatory subject may be seen as simply another part of the regulatory environment: it may be the focus of the regulator's attentions, but it also part of a complex system, and as we saw when discussing the Gardener's Dilemma, the subject of the regulatory intervention affects the regulatory environment as much as potential regulators. Thus we can remodel Figure 2.7 to reflect environmental realities. In so doing we see a complex regulatory environment emerge:

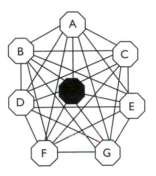

Figure 2.8 Information in China: True regulatory model (simplified).

What we see is a complex regulatory web, where the actions of each of our seven identified regulators (points A-G on the model) effect not only our regulatory object, which sits at the centre of the web, but also each other. Thus any regulatory environment forms part of a complex system, meaning that one cannot simply alter any one regulatory modality and hope to capture the regulatory object. All regulatory settlements, in the physical and virtual environments are a form of regulatory equilibrium achieved between all those who make up the regulatory environment. Changing any aspect of that regulatory equilibrium can have unexpected consequences.

101 Lessig, see above, fn 23, p 86.

Complex systems, layers and regulatory webs

As all regulatory environments can be defined as complex systems, regulatory settlements may, as we have seen, be defined as a balance achieved between complex regulatory components. Cyberspace, with its uniquely malleable environment and its stratified or layered structure is a particularly complex regulatory environment, meaning that mapping or forecasting regulatory settlements within cyberspace is especially difficult. Many computer scientists have turned to systems theory to attempt to map the complex structures of the network environment,[102] yet to date relatively few commentators on cyber-regulation have done similarly.[103] As we saw above[104] it is impossible to estimate the effects of any change, however minor, in the regulatory frame- work of any regulatory subject: a problem known in Chaos Theory as the 'butterfly effect'.[105] Thus any change at any point in the regulatory web can have immeasurable repercussions. If we return to our earlier case study exam- ining the regulation of media content in China we can see the effect of this problem. If the government of the PRC were to interfere in the developing market for information by nationalising and taking control of all Chinese media outlets it may imagine it would regain control of informational content in the PRC. This though would only occur should there be no *external effects* of this move: that is if all other regulators remained passive and unaffected. It is more likely though that such actions would cause unforeseeable responses from the other regulators. For example, in response the US International Broadcasting Bureau may increase output in Chinese and strengthen systems developed to circumvent state controls, or perhaps the market for informa- tion would simply be driven underground, increasing the black market for uncontrolled information supplied through technologies such as satellite, SMS or internet newsgroups. It is impossible to state what the effect would be due to the computational problems caused by complex systems. Thus no regulator can ever predict with certainty that the regulatory action they take

---

102   Many such papers can be found in the Journal *Theory of Computing Systems* published six times per annum by Springer-Verlag New York, LLC.

103   The most obvious exception being David Post and David Johnson, who in 1998 produced a detailed paper looking at chaos theory and cyber-regulation: 'Chaos Prevailing on every Continent: Towards a New Theory of Decentralized Decision-making in Complex Systems', see above, fn 12.

104   *Regulatory Competition, Regulatory Modalities and Systems*, see above, p 23.

105   The Butterfly Effect is prevalent in Chaos Theory. It is used to model how minute changes can have huge cumulative consequences in complex systems: thus the flapping of a single butterfly's wing today produces a tiny change in the state of the atmosphere. Over a period of time, what the atmosphere actually does diverges from what it would have done. So, in a month's time, a tornado that would have devastated the Indonesian coast doesn't happen. Or maybe one that wasn't going to happen, does. See Stewart, I, *Does God Play Dice? The Mathematics of Chaos*, 1990, Oxford: Blackwell, p 141.

will have the effect they desire. This, of course, does not mean that all attempts at regulation are doomed to failure; it simply means that whenever a regulatory intervention occurs it will have some unexpected or unpredictable results. In most cases these unexpected side effects are minimal and do not affect the effectiveness of the regulatory intervention and in the physical world such problems are reduced by the predictability of environmental regulators, which must comply with the constant universal regulator of the laws of physics. In the physical world, changes that occur in socially mediated modalities of regulation will, due to environmental inertia, not generally impact upon environmental modalities. This means in calculating the means to achieve a desired regulatory settlement in the physical world, the response of environmental modalities may be accurately predicted. Thus any regulatory intervention need only predict the responses of the socially mediated modalities: laws, markets and norms. Further, as most interventions are carried out by states or state-sanctioned regulators, law may frequently be quite accurately modelled as they have a high degree of transparency and flexibility. This means that although a regulatory intervention in the physical world can never be modelled completely accurately, regulators may reduce their exposure to unexpected side effects. As we have seen though, cyberspace is quite different. First, the physical laws do not apply to the higher layers (the logical infrastructure layer and content layer) and thus the environmental response to any regulatory intervention cannot be accurately modelled. Second, the effect of layering itself is to create an exponentially complex system with each added layer. If we return to our simple regulatory web seen in Figure 2.8, we see a flat or two-dimensional regulatory structure. This is a simplification of the true picture, one where social layering increases the complexity of the true regulatory picture. In cyberspace there is the added complexity of physical or environmental layering. Our simple two-dimensional web becomes a complex three-dimensional matrix (see Figure 2.9, page 54).

At each point in the matrix, a regulatory intervention may be made, but the complexity of the matrix means that it is impossible to predict the response of any other point in the matrix. Regulation within the complex, malleable, layered environment of cyberspace *is* considerably more complex to model than regulation within physical space. This complexity is exponentially more difficult to calculate with each added modality, and as a potential modality may be something as simple as a new software applet we see that cyberspace is becoming increasingly difficult to model.

Does this mean that cyberspace is unregulable? Were the cyber-libertarians right? The answer to both these questions is no. As the cyber-paternalists have conclusively proven, cyberspace is not only regulable, it is, due to the malleability of its environment, highly susceptible to regulation. Equally, though, it is difficult to predict the effects of a regulatory intervention in cyberspace. The introduction of a change in network protocols, or in legal rules governing aspects of cyber-society may have far-reaching and as yet,

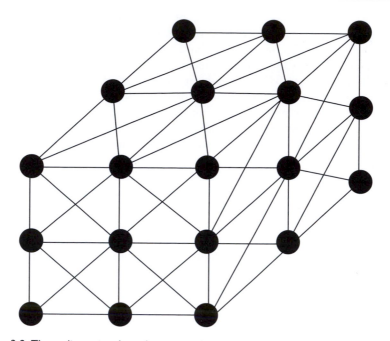

Figure 2.9 Three-dimensional regulatory matrix.

incalculable consequences. The following section of this book will examine the current regulatory environment within cyberspace. In five chapters, loosely organised along the lines of Lessig's four modalities, I will examine the current regulatory settlement. In doing so I hope to highlight the problems regulators have encountered in following a piecemeal approach to cyber-regulation. In the final section, I hope to demonstrate that what is needed is an approach that takes account of the unique nature of the network environment, a need for a more cohesive, measured, prudent and non-interventionist approach.

Part II

Regulatory tools and digital content

Chapter 3

Environmental design and control

In the beginning ARPA created the ARPANET
And the ARPANET was without form and void
And darkness was upon the deep
And the spirit of ARPA moved upon the face of the network
And ARPA said 'let there be protocol' and there was a protocol
And ARPA saw that it was good
And ARPA said 'let there be more protocols' and it was so
And ARPA saw that it was good
And ARPA said let there be more networks
And it was so.

Danny Cohen

The structure of modern cyberspace was set out by a group of computer visionaries in the 1960s. A group of men clustered around the leadership, and funding, of Professor JCR Licklider, the first director of the Information Processing Techniques Office (IPTO) at the Advanced Research Projects Agency (ARPA). This group included Bob Taylor, Wes Clark, Larry Roberts and Leonard Kleinrock – the men who set out systems and protocols for the management and distribution of information, which still form the backbone of the internet today. Reviewing the work of these pioneers we can identify two aspects of the digital environment which have been 'hardwired' into the network from its earliest inception: plasticity and layering. The first of these, environmental plasticity, is seen in the earliest work of Professor Licklider,[1] but is crystallised in his groundbreaking 1968 paper, *The Computer as a Communication Device*,[2] which he co-wrote with Bob Taylor. Here the

---

1   In his famous paper 'Man-Computer Symbiosis', *IRE Transactions on Human Factors in Electronics*, vol. HFE-1, p 4, March 1960, he developed the idea that computers should be developed with the goal 'to enable men and computers to cooperate in making decisions and controlling complex situations without inflexible dependence on predetermined programs'. (at p 4). At this time the idea of 'flexibility of machines' was groundbreaking.
2   *Science and Technology*, April 1968, p 21.

authors describe the programmed digital computer as 'a plastic or moldable medium that can be modeled, a dynamic medium in which premises will flow into consequences, a common medium that can be contributed to and experimented with by all'.[3] As outlined in Chapter 2, this plasticity survives in the modern network and is the key to understanding regulatory settlements within the modern digital environment. Cyberspace is the only social environment that is completely within human control. As a result, the effectiveness, and therefore role, of environmental, or design-based regulatory tools are increased.

Network layering may equally be identified in the earliest network designs. One of the first problems faced by the fledgling ARPANET[4] was how to make all computers on the network compatible with one another. In the 1960s there was no standard operating system such as Windows: each computer used a different programming language and operating system. Just to get the individual machines to interface, or 'talk', with each other was going to consume vast amounts of the ARPANET project's time and resources. At a meeting for principal investigators for the project at Ann Arbour, Michigan in April 1967, Wesley Clark, an engineer at Washington University, offered a solution. Instead of connecting each machine directly to the network they could install a minicomputer called an 'interface message processor', or IMP, at each site. The IMP would handle the interface between the host computer and the ARPANET network. This meant each site would only have to write one interface: that between that host and the IMP and, as all the IMPs used the same programming language, the network of IMPs would handle the rest.[5] Following this meeting, Larry Roberts, the IPTO chief scientist, prepared a paper for the Association for Computer Machinery (ACM) Symposium on operating system principles at Gatlinburg, Tennessee.[6] This paper, entitled *Multiple Computer Networks and Intercomputer Communication*,[7] set out Roberts' design for a layered network where access and information transfer would be managed by IMPs and processing would be carried out by the host supercomputers. The importance of these two 'network architectures' cannot be understated. The first means that the researcher should take a wider approach than the socio-legal approach usually followed, employing a socio-technological-legal, or STL approach, which encompasses

3   *Science and Technology*, April 1968, above, fn 2.
4   ARPANET is the forerunner of the modern internet. It will be discussed in greater depth below.
5   This is recorded by Hafner, K and Lyon, M, *Where Wizards Stay Up Late: The Origins of the Internet*, 1996, New York: Touchstone, p 73.
6   This symposium was held in October 1967 and is famous for not only being the place where Roberts first set out his layered design, but also for facilitating the meeting of Roberts and Roger Scantlebury, which will be discussed below.
7   Available from Larry Roberts' website at *www.packet.cc/files/multi-net-inter-comm.html*.

an examination of the role of technology in shaping and regulating our social environment. The second directs the researcher to be aware of the three-dimensional nature of the regulatory environment within cyberspace. This chapter will examine the role and effectiveness of design-based modalities in cyberspace against this background.

Mapping the digital environment

Before going on to discuss in detail the role played by harnessing design-based regulatory tools in the digital environment, it is essential to define what these terms mean, and what areas will be examined. The digital environment is a macro label, which may be attached to any area of human interaction that is facilitated by digital technology. This is a vast subject area encompassing: digital telecommunications (both fixed line and mobile); facsimile communication; digital broadcast media (both terrestrial and satellite); digital consumer products (such as DVD, digital photography and digital music); the internet; computer software; computer gaming; digital tracking and security products (such as radio frequency identification (RFID)); information management or databases; digital profiling tools (facial recognition technology and biometric profiling); and digital satellite technology (including GPS navigation). To analyse the regulatory framework surrounding all these products and services would be an overwhelming task. As a result, this chapter, and the following four chapters on network regulators, community controls, hierarchical regulation and indirect controls, will focus on only one aspect of this: the supply of digital content through networked environments. This means focusing on the traditional model of the internet as an area connected by wires, but also taking account of new digital network technologies such as third-generation mobile telecommunications: collectively, the wired and wireless networks that make up this place are known as cyberspace. In my analysis, I will focus upon the 'wired' internet. This is due to its rich history and regulatory framework; in doing so I will use the terms internet and cyberspace interchangeably.

The internet or cyberspace is an identifiable subset of the larger digital environment. It is defined as: 'the global information system that (i) is logically linked together by a globally unique address space based on the internet protocol (IP) or its subsequent extensions/follow-ons; (ii) is able to support communications using the Transmission Control Protocol/Internet Protocol (TCP/IP) suite or its subsequent extensions/follow-ons, and/or other IP-compatible protocols; and (iii) provides, uses or makes accessible, either publicly or privately, high-level services layered on the communication'.[8] In

---

8   This is the definition of the internet found in the Resolution of The Federal Networking Council of 24 October 1995. Here I use the term cyberspace as being synonymous with the

characterising cyberspace, it is impossible to define its boundaries by reference to what software suites or protocols it contains. This is because it is constantly evolving and over the years has consisted of computer networks such as Usenet, Milnet and MUDs (or multi-user dungeons); communications clients such as IRC (Internet Relay Chat) and email; and more recently macro-networks and multifunctional networks such as the World Wide Web (WWW) and peer-to-peer (P2P) communications. While most people equate the 'internet' with the web, and many believe email is separate from the internet environment, neither of these are true. Both these, and other systems such as P2P, WAP and Instant Messenger all travel over TCP/IP and the true definition of cyberspace is to be found in this powerful protocol. The difficulty in defining an environment by reference to a suite of communication protocols is that those who are not familiar with the technology will fail to understand exactly where the borders of this place are to be drawn. To help define cyberspace in a more user-friendly way, therefore, I will begin by revisiting the history of the place. By examining how cyberspace came about I hope to make clear where it may be found. This means returning to the work of Professor Licklider and the IPTO.

The history of the internet: ARPANET

The number of histories of the internet that have been published, both online and offline are too great to list. I do not intend to examine the history of the internet in detail. Those who wish to learn in detail about the development of this fascinating place should either read the excellent *Where Wizards Stay Up Late* by Katie Hafner and Matthew Lyon[9] or the Internet Society's *A Brief History of the Internet*,[10] which is without doubt the best online history of cyberspace. Much of the following is drawn from these two sources.

The starting point for any discussion of the origins of the internet should be to dispel a popularly held urban myth. It is often recounted, frequently by people who should know better, that the internet was designed for the Pentagon to provide a secure communications network, which would survive a direct nuclear strike. The reason this myth persists is due to a shared history between the internet and other work being carried out for the Pentagon at the time by Paul Baran of the Rand Corporation. The true origins of the internet

internet: this is an acceptable use of the term within the technical community, which usually defines cyberspace as 'The notional environment within which electronic communication occurs, especially when represented as the inside of a computer system'. Source: *Oxford English Dictionary*.

9 See above, fn 5.
10 Available at: *www.isoc.org/internet/history/brief.shtml*. This is written by many of the key players at the time including Vint Cerf, Bob Kahn, Leonard Kleinrock, Bob Postel and Larry Roberts.

are rather more prosaic. On 4 October 1957 the Soviet Union launched the first man-made satellite, Sputnik I, beating the United States into space. This event, which had not been predicted in the West, caused immense shock and surprise to the US scientific establishment. President Eisenhower was determined to ensure that the US would never again be taken by surprise on the technological frontier. In an immediate response to this event he created a new research agency tied directly to the Office of the President and funded from the Department of Defense budget. This new agency would oversee cutting-edge research of value to both the civilian and military establishments. It was to be called the Advanced Research Projects Agency or ARPA and, as if to underline its civilian credentials, it would be under the control of Roy Johnson, previously an executive with General Electric.[11] One of the first problems put to ARPA was how to deal with the inefficient use of expensive scientific equipment. One particular problem was that at that time computers were expensive pieces of equipment that were underutilised. They used batch processing techniques, which meant that hours or days could be spent inputting data on punch cards before the program could be run; then the slightest error in data entry would invalidate all this work. In-between time the computers often lay idle.

By the early 1960s every computer scientist wanted his/her own computer, but the cost of providing this was prohibitive. The answer was clear: the users had to share the resources available more efficiently. This meant two things: first, the development of time-sharing mainframe resources, an idea first put forward by researchers at MIT's Lincoln Lab in the 1950s,[12] and second, a network of machines that would allow researchers in different parts of the country to share results and resources easily. This idea was rooted in JCR Licklider and Wes Clark's August 1962 paper, *On-Line Man Computer Communication*. In this they described a 'Galactic Network', which 'encompasses distributed social interactions through computer networks'.[13] It was in part-response to this paper that Licklider found himself appointed as the first project director of the IPTO in October 1962. Immediately following his appointment, Licklider and his team set to work in developing a network technology that would give effect to his vision. Led by researchers such as Bob Taylor, they started work on making a shared computer network a reality. One of the key players in this was Leonard Kleinrock of MIT, who in

11 ARPA came into existence on 7 February 1958. Available at *www.arpa.mil/body/arpa_darpa.html*.
12 Hafner and Lyon, see above, fn 5, p 25.
13 The paper was originally published as an ARPA memo, one of a series by Licklider on this subject in 1962. Reprinted in Proceedings of the IEEE, Special Issue on Packet Communications Networks, vol. 66 no.11, November 1978.

1961 had published the first paper on packet-switching theory.[14] Kleinrock convinced Taylor that the best way to connect computers on a network was by using packet-switching rather than conventional circuit switching.[15] In 1966, when Taylor was promoted to Licklider's old post, one of his first actions in his new role was to hire Lawrence Roberts as IPTO chief scientist with particular responsibility for building a computer network, now to be called the Advanced Research Projects Agency NETwork or ARPANET. Roberts set to work developing his plan for the ARPANET, but found it difficult to design a functioning packet-switched network. At the ACM Symposium at Gatlinburg, Tennessee in October 1967, Roberts presented his first paper on the design of the ARPANET, the aforementioned *Multiple Computer Networks and Intercomputer Communication*,[16] but he had still to solve the packet-switching problem. He chose instead to close his presentation with a discussion of 'communication needs'. He noted that phone lines were slow and keeping a circuit (or line) open was wasteful, but he admitted he hadn't as yet found a more efficient method of transmitting data.[17] Amazingly, at the same symposium, Roger Scantlebury, an engineer from the National Physical Laboratory (NPL) in Teddington, London, was delivering a paper that gave a detailed outline of how to build a packet-switching network. After delivering his paper, Scantlebury outlined to Roberts how, using the system he and Donald Davies had developed at the NPL, he could design ARPANET to carry data at 20 times the speed of his original outline. Scantlebury also told Roberts that ARPA had funded some early research into packet-switching between 1960 and 1962 – research carried out by Paul Baran at the Rand Corporation. ARPA had commissioned Rand to design a voice communications network that would continue to function after a nuclear strike. Baran's solution was to harness packet-switching technology and distributed networks.[18] In 11 Rand studies published in 1962, which were later

14 Kleinrock, L, 'Information Flow in Large Communication Nets', *RLE Quarterly Progress Report*, July 1961.
15 Circuit-switching occurs when a dedicated channel (or circuit) is established for the duration of a transmission. All information is transmitted along this one dedicated link. The most ubiquitous circuit-switching network is the telephone system, which links together wire segments to create a single unbroken line for each telephone call. Conversely, packet-switching occurs when messages are divided into segments (or packets) before they are sent. Each packet is then transmitted individually and each can follow a different route to its destination. Once all the packets forming a message arrive at the destination, they are recompiled to form the original message. Packet switching is generally seen as more efficient as there is no 'silent time' between sections of the message as there is with circuit-switching.
16 Above n 7.   17 Hafner and Lyon, see above, fn 5, p 76.
18 A distributed communications network has built-in redundancy, which gives it greater resilience. It is one of three network models. The most basic is a centralised network, an example of which is a mainframe computer, which may run several 'slaves'. Here, all nodes are connected directly and only to a centralised hub or switch. All data is sent from an individual node to the centre and then routed to its destination. If the centre is destroyed or not

gathered together in the 1964 publication *On Distributed Communications*,[19] Baran outlined the necessary designs for a distributed communications network that would harness packet-switching. Unfortunately, these reports had been gathering dust in the IPTO archives until Scantlebury brought them to Roberts' attention.[20] It was from Baran's Rand study that the urban myth arose. It claims that ARPANET was part of a military communications network designed to withstand a nuclear first strike. This was never true of the ARPANET, only the unrelated Rand study on secure voice communications. However, later work on internetworks did emphasise robustness and survivability, including the capability to withstand losses of large portions of the underlying networks. With Baran's work now available to him, Roberts, with partners at Bolt, Beranek & Newman, a small computer company in Cambridge, Massachusetts, started building the network. Finally, on 29 October 1969, the ARPANET vision became reality when Charlie Kline, an undergraduate at UCLA, successfully logged in to the SDS 940 host at the Stanford Research Institute through the Sigma 7 host at UCLA.[21] Further hosts were added at the University of California at Santa Barbara and at the University of Utah, and by December 1969 the now familiar original four-node network seen in Figure 3.1 was in place.

ARPANET may have been the first successful computer network, and it may also have been the forerunner of the modern internet, but it was quite dissimilar to the internet as we know it today. Modern definitions of the internet describe it as a 'vast collection of interconnected networks'.[22]

---

functioning, all communication is effectively cut off. If the route between a node and the centre is destroyed or not functioning, that node is effectively cut off. Second, there is a decentralised network. The traditional telephone network is an example of such a network. It uses several centralised hubs, like several small centralised networks joined together. Each individual node is still dependent upon the proper functioning of its hub and the route to it. A distributed network is quite different. Such a network has no centralised switch. Each node would be connected to several of its neighbouring nodes in a lattice (or web) configuration. This means each node has several possible routes to send data. If one route or neighbouring node is destroyed, an alternative path would be available.

19 The studies may be accessed at: *www.rand.org/publications/RM/baran.list.html*.
20 Baran's work at ARPA despite 'gathering dust' for five years now forms the foundation of almost all new communications technologies and has garnered him several awards including the 1990 IEEE Alexander Graham Bell Medal awarded for 'Pioneering in Packet Switching' and in 1991, he received the Marconi Prize for the original ideas underlying packet-switching and many other communications advances.
21 Commentators, including Leonard Kleinrock, recall that the first attempt by Klein to Login failed when the network crashed halfway through the procedure – see Kleinrock, L, *The Birth of the Internet*, 1996, available at: *www.lk.cs.ucla.edu/LK/Inet/birth.html*. Later that same day though, Klein succeeded. The actual log entry recording the successful connection may be seen at: *www.computerhistory.org/exhibits/internet_history/full_size_images/imp_log.jpg*.
22 This definition is taken from Google, but it reflects the generally accepted view stated elsewhere.

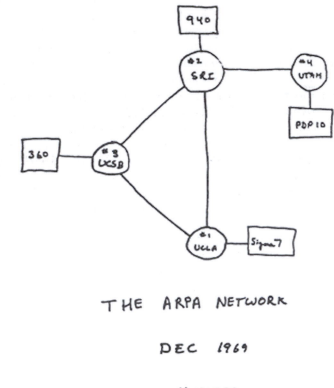

THE ARPA NETWORK

DEC 1969

4 NODES

Figure 3.1 Contemporary drawing recording the original ARPAnet network.

ARPANET was a single network. Furthermore, ARPANET was what today would be described as a 'closed network': you could only gain access to the ARPANET network if you had a correctly configured IMP, which had to be supplied by Bolt, Beranek & Newman. Thus the foundations of the modern internet architecture are not to be found in this part of the network's history. What happened next though was something quite special.

The history of the internet: TCP/IP

The ARPANET success was an exciting moment for network engineers. Bob Taylor, Lawrence Roberts, Robert Kleinrock and the extensive backroom team at Bolt, Beranek & Newman had demonstrated that functioning computer networks could be constructed. As a result the early 1970s were a time of intense experimentation with computer networking and other applications of packet-switching. One of the first experiments was in the carrier medium.

ARPANET used the existing AT&T telecoms system to carry its messages, but this was less efficient in areas where there was a lack of telecoms coverage: areas such as inter-island communications in Hawaii. In 1969, one of Bob Taylor's final acts as project director at the IPTO was to award funding to Professor Norm Abramson at the University of Hawaii to develop a wireless network. Abramson used this money to construct a simple network of seven computers across the islands using radios similar to those used by taxis to transmit and receive data. This network, called ALOHANET, used a different transmission system to ARPANET. In ARPANET communications the IMPs would manage data transmission and reception ensuring data was properly sent and received without interference. In ALOHANET, the terminals were allowed to transmit whenever they wanted to; if the transmission was impeded by other traffic the recipient computer would ask for it to be resent; the sending computer would keep sending the message at random intervals until it got an 'ok' message from the recipient. ALOHANET received a lot of attention, not least from the military, who recognised the advantages of a wireless network. The problem was, though, that the range of the network was limited and to build larger transmitters would centralise the network leaving it open to attack – the very problem Baran had wrestled with 10 years before. An alternative was to use satellites which were becoming more common at the time. Although slower than ARPANET, due to transmission lag, a satellite network would allow for international transmissions,[23] and for a degree of mobility in the network. To this end the US, the UK and Norway collaborated on the development of another network – SATNET, designed around satellite communication. Concurrently, local fixed-line networks were being developed in the UK, where Donald Davis at the NPL was developing a packet-switched network, and in France where Louis Pouzin had developed Cyclades, a network that put the burden of reliability on the host computers themselves, in much the same way as the ALOHANET did, by forcing host computers to ask recipients for acknowledgement of receipt of the data.[24]

With the development of several independent networks, interest in linking these resources grew. Bob Kahn, who had helped design the IMPs at Bolt, Beranek & Newman, was working on a packet-radio project at the time.[25] He wanted to connect his network to the ARPANET computer network, but at the time this was impossible as they were radically different networks. Kahn

---

23 At the time the fixed-line, or undersea, network, which was constructed of copper-wire, lacked the necessary capacity between the US and Europe to allow for effective network transmission between the two. This was remedied soon after when the telecoms companies laid high-speed fibre-optic cable in its place.

24 All this is discussed in Hafner and Lyon, see above, fn 5, ch 8.

25 Kahn was a professor of electrical engineering at MIT, who joined Bolt, Beranek & Newman to help them overcome communications errors in data transmission.

and others sought to end their frustration. What they wanted was a network of networks: an internetwork. They formed a group called the International Network Working Group (INWG) and appointed Vint Cerf to be its chair. That same year, 1972, Kahn was invited by Larry Roberts to join the IPTO to work on a net project: The Internetting Project. Kahn accepted and immediately started developing designs to connect together all the independent networks that had sprung up since 1969. Kahn realised the solution to the problem was in open architecture networking. Open architecture allows for each individual network to retain its unique network architecture, while connections between networks take place at a higher 'Internetworking Architecture Layer' as seen in Figure 3.2.[26] In an open architecture network, the individual networks may be separately designed and developed and each may have its own unique interface, which it may offer to users and/or other providers. Each network can be designed in accordance with the specific environment and user requirements of that network. There are generally no constraints on the types of network that can be included or on their geographic scope, although certain pragmatic considerations will dictate what it makes sense to offer. Designing this network was similar to designing the ARPANET. Like ARPANET the problem was that each host (in this case host network) used its own language. What was needed was a version of the IMP system that had bridged this gap on the ARPANET. Unfortunately, the IMPs were also one of Kahn's major obstacles. The language of ARPANET, the Network Control Protocol (NCP), did not have the ability to interface with networks or machines further downstream than a destination IMP on the ARPANET.[27] Thus Kahn would need to rewrite the NCP protocol.

Kahn therefore set about designing a new open architecture protocol. In doing so he set out four ground rules for his new internetwork protocol: (1) Each distinct network would have to stand on its own and no internal changes could be required to any such network to connect it to the internet; (2) Communications would be on a best-effort basis. If a packet didn't make

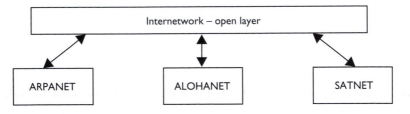

Figure 3.2 Simplified open architecture network.

---

26 This is set out more fully in *A Brief History of the Internet*, see above, fn 10.
27 The reason for this short-sightedness was that in designing NCP control over packets of data, network reliability was given to the IMPs. This is because this was simpler and cheaper and at the time ARPANET was the only network envisaged by the designers.

it to the final destination, it would shortly be retransmitted from the source; (3) Black boxes would be used to connect the networks (these would later be called gateways and routers). There would be no information retained by the black boxes about the individual flows of packets passing through them, thereby keeping them simple and avoiding complicated adaptation and recovery from various failure modes; and (4) There would be no global control at the operations level.[28] Kahn began working on his new protocol with his old friend and colleague, Vint Cerf, who joined in his role as chair of the INWG. According to Cerf, Kahn introduced the problem to him by saying: 'Look, my problem is how I get a computer that's on a satellite net and a computer on a radio net and a computer on the ARPANET to communicate uniformly with each other without realising what's going on in between.'[29] Cerf was fascinated by the problem and along with Kahn he worked on developing the new protocol, which would allow transmission across networks. The problem was each of these networks had its own set of rules. They used different interfaces, different transmission rates and each allowed for differently sized packets of information to be carried. How could they write a protocol that could be used uniformly across networks? Throughout the summer of 1973 Cerf and Kahn continued to work on these problems. Finally, in September 1973, Cerf presented his and Kahn's ideas at the INWG meeting at the University of Sussex. Their idea, which was later refined and published as their seminal paper, *A Protocol for Packet Network Intercommunication*,[30] was deceptively simple. Cerf realised that carriers often carried packets without ever knowing what was in them. A transport container has a standard size and shape, yet it may be carrying anything from video recorders to bikinis. Due to the common size of the container it can be carried by road, sea or rail and neither the ship's captain or the truck or train driver need know what he is carrying. The only people who need know are the shipper and the recipient. Cerf applied this to digital data. He designed a new protocol: transmission control protocol, or TCP, which would 'box up' the information and address it. Each message fragment or datagram would be the same size and could be handled by any of the networks; equally, once sent into the network, packets could take any route to their destination; they were not all destined to follow each other across busy networks. This design would work, and allowed for transmission of data across networks, but by removing the IMPs from the process, there remained the problem of missing or damaged datagrams. In ARPANET, the IMPs were responsible for sending and reassembling all message packets. They worked to ensure message integrity by

28 Taken from *A Brief History of the Internet*, see above, fn 10.
29 Hafner and Lyon, see above, fn 5, p 223.
30 Cerf, V and Kahn, R, 'A Protocol for Packet Network Interconnection', *IEEE Trans on Comms*, vol. Com-22, No 5, 627, May 1974.

checking the message at every stage of its journey – so-called hop-by-hop transmission. Cerf and Kahn changed all this in designing TCP. They returned to Louis Pouzin's work with Cyclades and Norm Abramson's work on ALOHANET. To overcome the problem of interference or data corruption both these networks had placed the responsibility for data integrity on the sending and receiving computers – so-called end-to-end reliability. In Cerf and Kahn's TCP design, when packets of information were sent they carried with them a request for acknowledgement. If safely received the recipient host would signal the transmission host of success. If the packet failed to arrive or was corrupted in transmission the recipient would not signal. If no acknowledgement was received the transmission host would retransmit the packet at random intervals until a successful acknowledgement was received. By placing all these responsibilities with the hosts, the network itself could be significantly simplified. Like the container transports of the real world, only the sender and recipient need know the details of the contents. All the network needed to know was where to send them.

TCP was not quite an instant success. Although it did lead to the development of the first internet, a network which between 1973 and 1975 grew at a rate of about one new network node per month,[31] the protocol was redrafted and redeveloped continually over the next few years. The most important of these occurred in January 1978 when Vint Cerf and Jon Postel posted IEN 21,[32] *TCP Version 3 Specification*, which suggested the splitting of TCP into a dual protocol TCP/IP. TCP had always been a multifunctional protocol, but by splitting it into the dual layer TCP/IP, these functions could now be clearly seen. The TCP element of the protocol breaks the data into packets ready for transmission and recombines them on the receiving end. The IP handles the addressing and routing of the data and makes sure it gets sent to the proper destination. Despite these developments, it was still not clear that the new TCP/IP protocol would continue to form the basis of the inter-networking layer. A rival protocol existed. OSI, or open-system interconnection, had been developed by the International Organisation for Standardisation (ISO). The ISO didn't believe TCP/IP was the best answer for the internet: they categorised it simply as an academic experiment that had outlived its usefulness.[33] They believed OSI was a better, more robust, protocol. Among the ARPANET community though, TCP/IP had the support of users. They felt OSI was too compartmentalised and complex. Further, OSI was just a design,

---

31 Hafner and Lyon, see above, fn 5, p 228.
32 The IEN or Internet Experiment Note series were a series of reports pertinent to the development of the internet. They formed part of the informal 'peer review' process common to the development of all computer networks and protocols. IENs were published in parallel to Requests for Comments (RFCs) which are the primary reporting process. IENs are no longer active
33 Hafner and Lyon, see above, fn 5, p 247.

TCP/IP worked. On 1 January 1983, TCP/IP took a major step towards acceptance when the ARPANET network made TCP/IP its operating protocol. This was a major development for the fledgling internet as it meant that due to the open architecture of TCP/IP the network could branch out in any direction. Vint Cerf joked that, '[t]o borrow a phrase, now it could go where no network had gone before'.[34] Despite this development, the ISO continued to develop the OSI protocol and beginning in 1986 they began to publish details of the new OSI standards.[35] Once these details were finalised, in 1988, the US Government announced their intention to immediately adopt the OSI protocol as the network protocol of the civilian ARPANET,[36] now known simply as 'the internet'.[37] Despite such popular support for OSI from government agencies in the US and elsewhere, support it still receives, it failed to replace TCP/IP. The reason for its failure is that during the period OSI was being developed, TCP/IP was being used. By 1989, virtually everybody was using TCP/IP and increasingly large infrastructures had been built upon it. Rebuilding networks around a new protocol was simply unthinkable. Through quiet momentum and granular development TCP/IP prevailed over

34 Quoted in Hafner and Lyon, see above, fn 5, p 249.
35 Publication took place in a series of ISO documents. The key document was ISO 8602:1987 *Information Processing Systems – Open Systems Interconnection – Protocol for Providing the Connectionless-mode Transport Service*; this was supported by ISO 8649:1988 *Information Processing Systems – Open Systems Interconnection – Service Definition for the Association Control Service Element*; ISO 8650 1988 *Information Processing Systems – Open Systems Interconnection – Protocol Specification for the Association Control Service Element* and at least a dozen other ISO specifications. All the original specifications have now been withdrawn and replaced with new specifications promulgated in the mid–1990s.
36 In January 1988 it was announced by the Department of Defense in RFC 1039, *DoD Statement on Open Systems Interconnection Protocols*, that 'The OSI message handling and file transfer protocols, together with their underlying protocols as defined in GOSIP, are adopted as experimental co-standards to the DoD protocols which provide similar services'. Although this language may not be clear to non-experts, it does signify adoption of OSI standards within the US Government, a position formally adopted in RFC 1169, *Explaining the Role of GOSIP*, where it was said that, 'Beginning in August, 1990, federal agencies procuring computer network products and services must require that those products and services comply with OSI, where feasible, as specified in GOSIP Version 1.0.'.
37 In 1983, once ARPANET switched to the TCP/IP protocol the Defense Communications Agency decided ARPANET was now too large (and open) to ensure security for the military ARPANET nodes. They, therefore, took the decision to split the ARPANET into two parts: Milnet for sites carrying non-classified military information and ARPANET for the computer research community. The split took 68 nodes into Milnet and left 45 in ARPANET. Gateways between the two remained and with TCP/IP in place most users didn't notice any difference. Then in 1984 the US National Science Federation began to construct a backbone of high-speed supercomputers for the computer science community. The network was launched in 1986 and was connected to ARPANET through TCP/IP. By 1988 the term ARPANET was fast becoming obsolete and people would generally use the term 'internet', although for many computer scientists this in effect meant NSFnet.

the OSI standard. Its success reflects much of the early network development. TCP/IP had been developed by Vint Cerf and Bob Kahn in meetings in Kahn's office at ARPA and in the Hyatt hotel in Palo Alto. The initial design of the TCP protocol had been drawn up by Vint Cerf while sitting in the lobby of a San Francisco hotel in-between conference sessions. The IP division in the protocol later added by Vint Cerf, Jon Postel and Danny Cohen was developed during a discussion held in the break of a meeting to discuss TCP at the University of Southern California's Information Sciences Institute. On that occasion, Jon Postel recalls that the ideas were developed by 'drawing diagrams on a big piece of cardboard that we leaned up against the wall in the hallway'.[38] By comparison the OSI standard had been developed by an intergovernmental agency of 40 years standing with over 100 members, more than 500 staff and an annual budget of over 100 million Swiss francs. As is recounted by another, anonymous, internet pioneer, the success of TCP/IP provides an object lesson in technology and how it advances. 'Standards should be discovered, not decreed.'[39] This tale is a valuable lesson for *anyone* who seeks to impose regulatory order within cyberspace.

The modern cyberspace

The modern internet, although developed from the original 1970s network, and founded upon principles of open architecture and the TCP/IP protocol, feels quite different for the average user. This is due to the astonishing pace of development that was seen in the 1990s. In 1990, ARPANET officially ceased to exist. It was replaced by the myriad of networks that had taken its place, the primary of which was NSFnet, a backbone network put in place by the US National Science Federation in the 1980s. The internet was slowly coming to life around academic networks such as NSFnet and, in the UK, JANET (the Joint Academic NETwork) based around 200 sites, which included higher education institutions, external research institutions, polytechnics and colleges. In October 1989 there were 159,000 internet hosts; by July 1999, this figure had risen to 56,218,000 hosts; today, according to the latest Internet Systems Consortium survey carried out in January 2006, there are 394,991,609 hosts.[40] The explosive growth of the network seen in the 1990s may be traced to two events, one a technological advance, the other a deregulatory action. Fortunately the two coincided, leading to the birth of the modern internet.

---

38 Quoted in Hafner and Lyon, see above, fn 5, p 236.
39 Quoted in Hafner and Lyon, see above, fn 5, p 254.
40 This data is drawn from the Internet Systems Consortium Domain Survey: Number of Internet Hosts, available at: *www.isc.org/index.pl?/ops/ds/*. All figures which follow are taken from this source.

The first was an invention. Tim Berners-Lee was a physicist who had graduated from the University of Oxford in 1976. Upon leaving university he began working with Plessey Telecommunications as a software engineer, where he worked for two years on distributed systems, message relays and bar coding. He then joined DG Nash, a small software company, where he developed a multitasking operating system and typesetting software for intelligent printers. During this time he also developed a hypertexting system called Enquire. Hypertext was not new – the term had been coined by filmmaker and computer programmer Ted Nelson in 1963 and the concept of hyper-linking was at the core of Project Xanadu, his hypertext project, which ran from 1960 and which he discussed at length in his 1981 book, *Literary Machines*.[41] While Nelson was experimenting with hypertext, the first functioning hypertext system, Douglas Englebert's NLS, or oN-Line System, was developed. Englebert developed his system independently of Nelson's work and did not use the term hypertext to describe his system, but there is no doubt it was the first hypertext network. Englebert had been hugely influenced by Vannevar Bush's 1945 paper *As We May Think*,[42] which described a mechanised library system, or memex, with embedded links between documents. In his attempts to build Bush's memex, Englebert turned to the potential of digital computers. In 1962 he started work on Augment, a project to develop computer tools to augment human capabilities.[43] This was possibly the most important computer project of the time (arguably even more important than ARPANET) and it produced the first computer mouse, graphical user interface (GUI) and hypertext program. All these developments were demonstrated by Englebert at the Fall Joint Computer Conference in San Francisco in December 1968. Englebert received a standing ovation and in tribute to his work, his NLS system, and in particular the SDS 940 computer he used for these applications, was selected as the second ARPANET node. It took another 20 years though for Englebert's invention to find a popular use, and that was in Tim Berners-Lee's web design.

After developing his Enquire system, Berners-Lee joined the European Particle Physics Laboratory (CERN)[44] as a consultant. During his time there he secured funding to develop a digital hypertext library of CERN research, which could be accessed from any facility on the CERN network. By March 1989, Berners-Lee had completed his project design to allow researchers in the High Energy Physics Department to communicate information online. His design had two key features: (1) like TCP/IP his new protocol was to have

41 Nelson, TH, *Literary Machines*, 93.1 edn, 1981, Sausalito, CA: Mindful Press.

42 Bush, V, 'As We May Think', *The Atlantic Monthly*, July 1945, p 101.

43 The Augmentation Research Center at Stanford Research Institute in Menlo Park, CA was the precursor to the internationally famous Xerox PARC facility.

44 CERN is the contraction of the Laboratory's French name Conseil Européen pour la Recherché Nucléaire.

an open architecture to allow researchers to connect any computer no matter what operating system it was using, and (2) information was to be distributed using the network itself, obviating the need for disks or CDs. Berners-Lee was joined in his project by Robert Cailliau, a computer engineer from Tongeren, Belgium. Throughout 1990 Berners-Lee, assisted by Cailliau, developed the first web server, 'httpd', and the first client, 'WorldWideWeb' a hypertext browser/editor. This work was started in October 1990, and by Christmas Day 1990 Berners-Lee and Cailliau were conversing across the world's first web server at info.cern.ch. In August 1991, Berners-Lee posted a notice to the alt.hypertext newsgroup informing users where his web server and browser software could be downloaded. At this stage the web was still in its infancy; there was no certainty it would develop in the way we experienced in the 1990s, but on 30 April 1993, the future of the web was secured when CERN gave notification that they were not intending to take control of the technology developed by Berners-Lee and Cailliau. On that date CERN announced and certified that the WWW technology developed at CERN was to be put into the public domain 'to further compatibility, common practices, and standards in networking and computer supported collaboration'.[45] This allowed any interested party to use and improve the CERN software, assuring the future of the web.

By April 1993 all the ingredients for the explosive growth of the internet were in place: a robust, open, network with a strong backbone provided by academic networks like NSFnet and JANET; a free and user-friendly protocol that allowed for easy access, retrieval and cataloguing of information; and an extensive growth in the number of installed home computers meaning there was a large latent market of users available. The final ingredient that was needed was the opportunity to profit from the network. Traditionally, profit-making activities had been banned from the network. The NSFnet acceptable use policy had always disallowed any commercial activity. This was to prevent valuable bandwidth from being redirected from the research work for which it was intended. By the late 1980s though, it was becoming clear that commercial interests wanted to use the network. To this end, Steve Wolff, then program director for NSFnet, decided to implement a network upgrade. Between 1988 and 1991 the NSFnet network was upgraded twice, increasing network capacity by more than 3,000 per cent. With the upgrades in place, the NSF amended its acceptable use policy in June 1992. Now commercial use by 'research arms of for-profit firms when engaged in open scholarly communication and research' was allowed; still not allowed was 'use for for-profit activities'.[46] Despite this, there was no holding back the

45 The original certificate may be viewed at: *http://info.web.cern.ch/info/Announcements/CERN/ 2003/04–30 Ten Years WWW/Declaration/Page1.html*.
46 The NSFNet acceptable use policy may be accessed at: *www.creighton.edu/nsfnet-aup.html*.

commercialisation of cyberspace and in 1994 the first commercial websites arrived.[47] In response the NSF began to seek commercial partners to take over the network backbone and during the same year four new network access points (NAP), operated by Sprint, Metro Fiber Systems, Ameritech and Pacific Bell, were put in place to take the strain off of the NSFnet. With the new commercial backbone in place the NSFnet was dissolved and the NSF returned to its roots providing a high-speed research network. With the deregulation of the network backbone, entrepreneurs began investing in cyberspace. Famous online business successes such as Amazon.com, CDNow.com and eBay.com,[48] along with famous failures Boo.com and Pets.com, all developed from the explosive cocktail of low start-up costs, almost unlimited growth potential and freedom from regulatory constructs that immediately followed deregulation of the network. This heady mix led directly to the dotcom bubble of the late 1990s, a bubble which at its peak saw the Bloomberg US Internet Index of 280 leading internet companies valued at $2.948 trillion dollars.[49] Even the bursting of this bubble has done little to extinguish the enthusiasm of entrepreneurs and investors for online businesses; hardly surprising given that the value of online business to consumer (B2C) commerce has increased from $38.8 billion in 2000, just before the dot com bubble burst, to $172 billion in 2005.[50]

To describe cyberspace as 'the internet' is therefore clearly rather simplistic. Cyberspace is an environment as rich as any other in which humans meet, talk, buy, sell and fall in love. It has a complex history and the foundations of modern cyberspace may be found partly in academic research facilities, partly in military research and partly in the commercial exploitation of the network, which followed its deregulation in 1994. This place is not only a network of networks, but is also a community of communities and as a result, designing environmentally based regulatory constructs is as difficult in cyberspace as in real space, as we shall see in the next chapter.

---

47 The first e-commerce website, as opposed to merely an advert placed on a Usenet discussion group, was set up by Pizza Hut in August 1994. This allowed customers to order pizza online (in a somewhat rudimentary fashion).

48 Amazon.com, CDNow.com and eBay.com were all formed in late 1994/early1995 and were talismanic successes in the dot.com boom.

49 Source: *The $1.7 trillion dot.com lesson*, CNN Money. Available at: *http://money.cnn.com/2000/11/09/technology/overview/*.

50 See Forrester Research (2005), *US eCommerce: 2005 To 2010: A Five-Year Forecast And Analysis of US Online Retail Sales*, 14 September 2005, available at: *www.forrester.com/Research/Document/Excerpt/0,7211,37626,00.html*.

Chapter 4

Code controls and controlling code

What a man's mind can create, man's character can control.

Thomas Edison

The network environment of cyberspace is an unplanned environment. It has, as we saw in the previous chapter, developed over the last 40 years in a piecemeal fashion, with most environmental changes designed to meet a particular demand, whether that be simple interconnectivity, efficiency, accessibility of content or commercial demands. Although the cyberpaternalists have demonstrated the efficiency of harnessing the power of the network environment as a regulatory tool,[1] there are considerable obstacles to be overcome before this can be achieved. Chief among these obstacles is the open architecture of the key network protocols, TCP/IP and HTTP.[2] By maintaining an open architecture it is difficult to create regulatory structures at either the internetworking (or TCP/IP) layer or the content (or HTTP) layer. These open designs help ensure that at the heart of the modern network retains Bob Kahn's rule that, 'there would be no global control at the operations level'. Thus to achieve regulatory control through environmental, or design, controls within the architecture of cyberspace, one has to identify the weak points within the network: the 'pinch points' where regulatory controls may be imposed and in so doing have the maximum effect on the largest possible number of internet users.

Pinch points can be found, most commonly, at the point of transition between layers, or networks, the original pinch point being the IMPs that connected the mainframe computers to the ARPANET. Thus the first point

---

1   See discussion above, pp 8–12. The Cyberpaternalist thesis is most fully explored in Lessig, L, *Code and Other Laws of Cyberspace*, 1999, New York: Basic Books.
2   HTTP, or hypertext transfer protocol, is the transmission protocol used by the World Wide Web. It was developed by Tim Berners-Lee for the CERN hypertext network and forms part of the WWW technology put into the public domain by CERN in 1993.

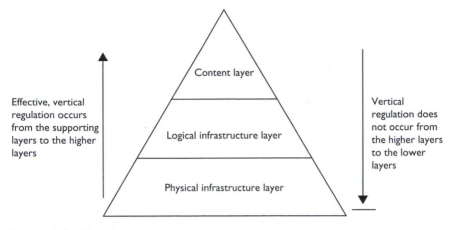

Figure 4.1 Benkler's layers.

of analysis for any student of cyber-regulatory theory is to return to Benkler's layers, and through these identify who controls the network environment at each layer, and how that person or persons may convert that environmental control into regulatory control. As was discussed in Chapter 2, the architecture of any communications network is stratified. Several models have been proposed to describe the network stratification seen in the internet, including the OSI seven-layer model used in all network design, Tim Berners-Lee's four-layer model used to describe the architecture of the web and Yochai Benkler's three-layer model.[3] Benkler's model, recreated in Figure 4.1 above, recognises three distinct network layers. The physical infrastructure layer is the basic foundational layer and comprises the necessary hardware to enable the network to function. It includes fixed line telecommunications networks, mobile telecommunications masts, and stations, cable networks, satellites, computer hardware, routers and gateways. The logical infrastructure layer sits atop this foundation. This is a connective, or internetworking, layer and comprises the network protocols TCP/IP, HTTP, SMTP, FTP, UMTS and many others.[4] Finally, at the top of this structure sits the content layer. It is a vast library of information and comprises all materials stored, transmitted and accessed using the software tools of the logical infrastructure layer.

The key to Benkler's layers, from the point of view of a regulatory theorist, is the relationship between the individual layers. As we discovered earlier it is only possible to vertically regulate across layers in one direction. This is

3   *Environmental Layers and Network Architecture*, see above, p 45.
4   A full listing and description of all TCP/IP, VoIP (voice over IP) and mobile protocols may be found at *www.protocols.com/*.

because the infrastructure of the network is like the physical infrastructure of a house. The physical infrastructure layer may be seen to equate to the foundations of the house. As a house must have solid foundations, the size, depth and strength of the foundations will dictate what size and shape of building may be erected. Similarly, the nature of the physical infrastructure has dictated to design of the network. The decision, to use packet-switching rather that traditional circuit-switching was dictated by the constraints of the copper-wire telecommunications network in place at the time. Similarly, the decision by the National Science Federation to disallow commercial activities on NSFnet, until after the network upgrade was completed in 1991, was determined by the available bandwidth within the NSFnet backbone. Thus the primary consideration of any regulatory action in relation to a designated environment, is the physical nature of that environment.[5] By altering the environment of the supporting layer, you may indirectly regulate activities in the higher layer(s).

Designing controls in the physical infrastructure layer

This is the central theme of Lawrence Lessig's second book, *The Future of Ideas: the fate of the commons in a connected world*.[6] In this he suggests that a key factor of control may be the leveraging of control from the physical infrastructure layer into the logical infrastructure and content layers. In particular, this book examined the threat such action may pose to creativity in the digital environment. Professor Lessig clearly believes in Stuart Brand's dictum that 'information wants to be free',[7] and in a carefully argued claim he

---

5   Environmental factors were key for example in siting military encampments and in particular castles. As noted by Sidney Toy: 'The castle is designed to be above all a stronghold, and is constructed on a variation of the standard medieval motte and bailey plan. In its simplest form, this consists of an earth mound, or motte, with some sort of palisade on top, sur-rounded by an outer wall, or bailey, creating or, more often, reinforcing the protection offered by the natural terrain – a protected space into which villagers or livestock could be gathered when threatened. At Durham, in northern England, the earth mound is combined with a naturally defensive position across the neck of a peninsula formed by the river Wear. The keep is now of stone, not wood (a technical development displayed, for obvious reasons, in most of the castles that survive to the present day) and the streets of North and South Bailey reveal the outline of the area which could once be enclosed and defended.' Toy, S, *Castles: Their Construction and History*, 1986, Minneola, NY: Dover Publications, p 135.

6   2001 New York: Random House.

7   Stuart Brand is one of the founders of Founded 'The WELL' (Whole Earth 'Lectronic Link), a computer teleconference system for the San Francisco Bay Area, one of the key network communities for intellectual discourse. He coined the phrase at the first Hackers' Conference, in autumn 1984. As Brand himself says in an email to TBTF, 'I said in one discussion session: "On the one hand information wants to be expensive, because it's so valuable. The right information in the right place just changes your life. On the other hand, information wants to

suggests that free resources are essential for creativity and innovation. This claim, it should be acknowledged, is part of a complex and ongoing discourse between supporters of the concept of a creative commons[8] in software, music, literature and film,[9] and those who support the extension of property rights into the digital realm,[10] and as such it goes beyond the scope of this book. Supporters of the intellectual commons ethos contend that a particular threat to the informational freedom currently enjoyed in cyberspace emanates from controls applied in the physical infrastructure layer, which may then be leveraged across both the logical infrastructure and content layers. This threat arises as a consequence of technological convergence, and the perceived effect of convergence on the supply of digital information. Technologists have long predicted that digital platforms will provide a cheap and simple cross-media delivery and carrier tool providing social, economic and political benefits.[11] The concept of convergence came to media and communications theory from the mathematical disciplines where it was used to refer to the coming together of physical things such as beams of light or non-parallel lines. Media and communications commentators began to apply the term to the coming together of media platforms in the late 1970s or early 1980s, it being extremely difficult to determine exactly when, and by whom, the term was first used in this context. What is clear though is that communications theorist Ithiel de Sola Pool adopted this contextual use of the term and popularised it among media and communications theorists. In his landmark 1983 book, *The Technologies of Freedom*,[12] Pool wrote of the 'convergence between historically separated modes of communication' and argued that 'electronic technology is bringing all modes of communications into one grand system'.[13] In this groundbreaking work Pool conjured up an image of a future where all content would be stored digitally, delivered over a network

be free, because the cost of getting it out is getting lower and lower all the time. So you have these two fighting against each other." That was printed in a report/transcript from the conference in the May 1985 *Whole Earth Review*, p 49.' This was also a major plank of both Richard Stallman and John Perry Barlow's attacks on the application of IP rights in digital information. See Denning, D, 'Concerning Hackers Who Break into Computer Systems', *Proceedings of the 13th National Computer Security Conference*, Washington DC, 1990, 653 (quoting Richard Stallman); Barlow, JP, 'The Economy of Ideas: Selling Wine Without Bottles on the Global Net', *Wired 2.03*, March 1994.

8 *http://creativecommons.org/*.
9 Supporters of the Creative Commons and Free and Open Source Software movements include Lawrence Lessig, Eric Eldred, Hal Abelson, James Boyle, Richard Stallman and Eben Moglen.
10 Supporters of Digital Property include Mark Stefik, Gerd Leonhard and Adam Thierer.
11 Eg Negroponte, N, *Being Digital*, 1995, Chatham: Hodder & Stoughton; de Sola Pool, I, *The Technologies of Freedom*, 1984, Cambridge, MA: Harvard UP (reprint); Brand, S, *The Media Lab: Inventing the Future at MIT*, 1987, Chicago, Donnelley & Sons.
12 Ibid.   13 Ibid, p 28.

and obtained through electronic devices. Pool's vision inspired many communications and media commentators to examine the propinquity and social impact of media convergence. Most assumed digitisation of platforms meant carrier convergence was near. In particular several key advances of the 1990s, including the emergence of the World Wide Web (www), the development of broadband connections, second-generation (digital) mobile telephony and the emergence of digital music and digital music platforms, caused commentators to predict the imminent convergence of platforms. In 1995, Nicholas Negroponte, in his landmark work on digitisation, *Being Digital*, confidently stated that: 'I am convinced that by the year 2005 Americans will spend more hours on the Internet (or whatever it is called) than watching network television.'[14] Despite Negroponte's prediction, the gap between the consumption of interactive online content and broadcast media remains as large as ever. A recent survey by Nielsen/Netratings showed that in February 2005, US web users averaged 13 hours and 44 minutes online, down 2 per cent from a year earlier,[15] while a series of reports, including the Kaiser Family Foundation report *Generation M*,[16] continue to report that the average US citizen watched in excess of 28 hours of television per week, or in excess of 121 hours per month. How could such a respected commentator as Nicholas Negroponte have been so far out in his prediction?

There are several reasons for the disparity between the predictions of media convergence made by eminent scholars such as Pool and Negroponte and the factual position today. One obvious reason is the failure to develop, as yet, a fully digital informational distribution chain. For complete digital convergence to become a reality we require technological innovations in every stage of the information infrastructure. Information needs to be gathered digitally. This is now becoming quite commonplace with reporters filing stories by email and the use of digital cameras to record news events and television broadcasts. Information then needs to be stored and delivered in digital form. Again this is slowly becoming true. Most media outlets now store information in digital form, but the distribution of media in digital form is less prevalent. Although digital television penetration rates are strong in the UK, it is still the case that one-third of UK households do not have access to digital broadcasts,[17] and it is by no means certain that Ofcom will be able to ensure sufficient market penetration to meet the 2011 date by which all UK analogue broadcasts (excepting the Channel Islands) are to be switched

---

14 Negroponte, see above fn 11, p 58.
15 Source Nielsen/Netratings Internet usage report, available at: *www.netratings.com/pr/ pr_050318.pdf*.
16 *Generation M: Media in the Lives of 8–18 Year-olds*, 9 March 2005, available at: *www.kff.org/ entmedia/entmedia030905pkg.cfm*.
17 The latest quarterly data from UK media regulator Ofcom, released in February 2006, revealed that 16.5 million homes, or 66 per cent, had digital TV as of September 2005.

off. Similarly online newspapers, although proving to be extremely popular are, on a whole, failing to overtake their paper counterparts. In April 2006, *Guardian Unlimited*, the online version of *The Guardian* received 3,704,050 unique visits from UK-based users, an average of 123,468 unique visits per day. For the same period the ABC circulation figure for the print version of *The Guardian* was 385,219 per issue, while its sister Sunday paper, *The Observer*, had circulation figures of 478,298 per issue.[18] Thus even the most popular of online newspapers has yet to usurp the paper version in the hearts of consumers. To overcome this we need to make digital information as accessible and portable as its paper-based counterpart. This requires first, a proliferation of wireless internet access points, something that although undoubtedly on the increase, is yet to have a substantial impact.[19] Second, it would require a new generation of portable devices that come closer to replicating the advantages of paper. Currently, mobile digital tools suffer either from portability or usability problems. To be small enough to be easily portable, and to have a reasonable battery life screens must be small, with the average screen size of a personal digital assistant (PDA) or mobile phone being less than four inches. This is too small to display clearly the density of text found in an average newspaper. To display the content in an easily access-ible fashion, something akin to a laptop computer must be used; these though are bulky, heavy and easily damaged, not the sort of thing one can read on a tube train. Thus, until digital carrier mediums can match the accessibility and portability of paper, digital distribution of news content will continue to lag behind traditional paper.[20]

Lawrence Lessig, like Ithiel de Sola Pool, Nicholas Negroponte and Cass Sunstein,[21] believes that convergence of media is close. In particular, when he wrote *The Future of Ideas*, which was published in 2001, he believed the model of this convergent action was at hand. On Monday 10 January 2000 two entertainment giants announced they were to merge. Time Warner was a traditional media organisation that described itself as one of the world's leading media companies. It owned the cable television channels CNN, TNT, Cartoon Network and Home Box Office, and was the publisher of magazines, books and websites. In addition, the company was active in the music business, produced feature films, television programmes and animation, and operated a cable TV business. At the time of the merger it employed 69,000 people worldwide and reported annual revenues of over $27 billion. America On-Line (AOL) was the world's largest internet service provider. It was founded in

18 Source *Guardian Unlimited*.
19 According to a recent Canadian report, wireless technology only provides 2 per cent of users with their primary network connection (mobile phone access excluded). Source Ipsos-Insight, *The Face of the Web 2004*.
20 You may apply the same rules to books and magazines.
21 *Republic.com*, Princeton: Princeton UP, 2001.

1985 to provide interactive services, internet technologies and e-commerce services. By 2000, the company had over 22 million customers and a market capitalisation of $163.4 billion. When these two industry giants announced their merger they said they were creating the world's first 'clicks and mortar' company.[22] At the time this was assumed to be the first of many such meg-amergers. Since then AOL/Time Warner have found it almost impossible to integrate the two arms of their company and they have posted some of the biggest losses in Wall Street history, including an overall loss for 2002 of $98.7 billion,[23] problems which have seen other companies stay clear of similar mergers. Despite this it is assumed that once market conditions are favourable there will be a return to such mergers and business models. For Professor Lessig, and others in the Creative Commons and Open Source movements, such vertical integration of carrier and content providers carries with it a particular danger.[24] This danger is that as technology platforms converge to provide a single platform technology a few companies take control of what becomes a single access point to the digital environment. The multiplatform environment we have today – television, radio, mobile tele-phone and computer with cable – will be replaced by a single converged access point with a single access protocol.

   In addition to the danger of convergent access, vertical integration of access providers and content providers, as seen in the AOL/Time Warner merger, may encourage discrimination in both carrier medium and content: AOL will favour content provided by Time Warner over other content, while Time Warner will provide content in a format that favours the AOL medium. Thus the controls that AOL can enforce in the physical infrastructure layer, leverages control in the content layer, favouring the preferred content pro-vider. For supporters of the creative commons though, the risk is more deeply seated than just creating business synergies. This is simply an example of how those who control the access platform control the logical infrastructure and content layers. For Lessig, there is a choice between two regulatory models:

> What's at stake here are two models for organizing a communications network, and the choice for us is which model will prevail. On the one

22 'Clicks and mortar' describes what had been to then the mythical union between internet growth and traditional levels of profit.
23 Source BBC News. See *http://news.bbc.co.uk/2/hi/business/2707199.stm*. Although it now appears the company has turned the corner with recent figures showing the group has made a profit of $1.5bn in the three months to March 2006. See *http://news.com.com/Time+Warner +profit+rises/2100–1014_3–6067761.html*.
24 Eg the work of the Center for Digital Democracy (*www.democraticmedia.org/*) and in par-ticular the views of its Director Jeff Chester as seen in his paper 'The Death Of The Internet: How Industry Intends To Kill The Net As We Know It'. Available at: *www.tompaine.com/ feature.cfm/ID/6600*.

hand, there is the model of the perfectly controlled cable provider – owning and controlling the physical, logical and content layers of its network. On the other hand, there is the model of the Internet – which exerts no control over a physical layer beyond the decision to include equipment or not, and which enables the free exchange of content over a code layer that remains open.[25]

He believes that as convergence and vertical integration continue to develop we will see cable companies, who control the physical infrastructure layer, taking control of more of the network:

> Here, then we have the beginnings of a classic 'tragedy of the commons'. For if keeping the network as a commons provides a benefit to all, yet closing individual links in the network provides a benefit to individuals, then by the logic that Garrett Hardin describes, we should expect the network 'naturally' to slide from dot.commons to dot.control. We should expect these private incentives for control to displace the public benefit for neutrality.[26]

Thus Lessig's argument is that there are two alternative regulatory models: the commons and the controlled, and that without intervention to protect the commons the natural evolution of cyberspace is for it to become a controlled environment because a controlled environment allows for profit-taking in a way a commons does not. Thus internet access providers such as cable and telecoms companies will design a closed environment. As media content converges this will give them control over audio-visual content including music, television and on-demand services such as movies on-demand and textual and multimedia content such as online newspapers, web pages, email and chat services. The open-source nature of the network is, in Lessig's terms, dependent upon the goodwill of the access providers who may change the nature of the place at any time. In support of this argument Lessig marshals two examples of control: one control exercised, the other control unutilised.

Professor Lessig's first example is of controls exercised by cable access providers. By referring to the work of Professor Jerome Saltzer of MIT,[27] he lists five examples of gatekeeping restrictions imposed by cable access providers on customers. The five are: (1) video limits – limiting the amount of minutes a customer may use streaming video; (2) server restrictions – failing to support, for example web hosting on the access provider's server; (3) fixed backbone – the long distance carrier is selected by the access provider;

---

25 Lessig, see above, fn 6, p 167.   26 Ibid p 168.
27 Saltzer, J (1999) ' "Open Access" is Just the Tip of the Iceberg', available at: *http://web.mit. edu/afs/athena.mit.edu/user/other/a/Saltzer/www/publications/openaccess.html.*

(4) filtering – filtering out packets including spam, pornography and shared files; and (5) restrictions on home networks – not allowing for home network connections. Both professors Saltzer and Lessig take these developments to indicate a leveraging of control from the physical infrastructure layer to the higher layers. Professor Saltzer suggests that server restrictions reflect a conflict of interest on the part of the access provider as they also offer a (separate) website hosting service, while filtering is dangerous as 'the access provider has an incentive to find a technical or political excuse to filter out services that compete with the entertainment or internet services it also offers'.[28] Both commentators are particularly concerned with video limits. Saltzer suggests another conflict of interests – 'they are restricting a service that will someday directly compete with Cable TV', while Lessig notes that 'cable companies make a lot of money streaming video to television sets'. The assumption both commentators make is that the cable access companies are making a positive intervention in a controlled environment. Both thus make the assumption of 'settled regulation' which, as seen previously, assumes that a regulator works within a settled environment and has time to positively consider policy decisions.[29] In truth, cable access providers are involved in a cut-throat marketplace with a high percentage of bankruptcies among providers. To assume that cable access providers who choose to put in place access restrictions to protect secondary markets in this way would be successful in the marketplace is irrational. The largest UK cable provider, NTL, has 3.14 million subscribers, just over 5 per cent of the UK population. Of these, 1,330,300 are broadband subscribers, meaning that it controls 21.78 per cent of the UK broadband market. This is less than market leader BT Broadband, which controls 24.41 per cent of the market and collectively cable companies NTL and Telewest provide just less than one-third of UK broadband access, the rest being provided by companies providing ADSL, or asymmetric digital subscriber lines, across the telephone network.[30] User restrictions, such as those described by professors Lessig and Saltzer have been a key strategy in the marketplace for broadband access in the UK, in particular, packages that impose download limits have been introduced by many ISPs to allow for a more competitively priced access package for low volume users. For example in the UK, market leaders BT Broadband provide monthly download restrictions on all their packages, ranging from a 2GB to a 40GB download restriction based upon the package bought. Competitors such as AOL, currently only the UK's fourth largest broadband access provider, provide unlimited downloads in an attempt to gain a market advantage over BT

---

28 See Saltzer, above, fn 27.
29 *Regulatory Competition, Regulatory Modalities and Systems*, see above, pp 23–30.
30 Source: UK broadband subscriber numbers and market shares, ZDNet, available at: *http:// blogs.zdnet.com/ITFacts/index.php?p=826*.

Broadband. Thus use restrictions are part of the market in broadband access. What we see is a complex market, with several competing service providers and a complex pricing and usage structure. As part of this market, code controls are utilised. Some suppliers such as BT offer lower prices but at a cost of download restrictions, others such as AOL offer a higher cost but with no restrictions. Attempts by cable access companies to leverage design controls to protect a secondary market in the 'streaming of video content', as described by professors Lessig and Saltzer, are unlikely to succeed in such a competitive primary market. What would result is a regulatory conflict between a design control imposed by a hierarchical order (the leveraged control of the cable providers) and a mature competitive market. In such a situation the mature market will, all other factors being neutral, 'route around' the attempts to leverage control unless they meet market conditions. Thus it is highly unlikely that any one operator, or group of operators, will, in a mature and competitive market, be able to leverage control from within the physical infrastructure layer to the higher layers. Such an action creates a market disadvantage in the primary market allowing the market to correct for such actions.

Professor Lessig's second example is one of a failure to regulate. He notes that when the ARPANET/internet network was in its infancy the telephone companies, and in particular the AT&T network managers, did not attempt to leverage the control they had over the wires to the higher carrier and content layers. Despite a substantial change in the way their wires were being used, and despite the potential for the new network to act in competition to the traditional telecommunications market, the management at AT&T 'stood by as the internet was served across their wires'.[31] He notes though that this result was not achieved simply from inertia or even charity on the part of AT&T: they were compelled to do so by the actions of the telecoms regulators.[32] The cause of this regulatory intervention, according to Professor Lessig, was disfavour with the current, monopolistic, telecoms supply model and that the intent of the regulator in breaking up this monopoly was to create competition in the telephone market. But, these actions had an unintended side effect: while regulators imagined they were creating competition in the supply of telephone services, they did not realise they were also allowing for the development of a product that would create competition with the telephone market. According to Professor Lessig, this achievement is now at risk. The freedom which the Federal Communications Commission (FCC) and network designers such as Vint Cerf and Bob Kahn created in the 1970s is at risk in the new millennium as cable access companies reclaim the controls that AT&T had in the 1970s. A new type of access monopoly allows

31 Lessig, see above, fn 6, p 148.   32 Ibid.

for control to be leveraged in the way that AT&T failed to do, or were prevented from doing, in the fledgling years of the network. In making this argument though, Professor Lessig makes a false assumption. He assumes that the market in internet access of the 1970s and the market of the new millennium are equally susceptible to monopoly control and capture. This is not the case. In the 1970s the market in internet access was an immature market. There was only one way to access the network: across the wires of established telecoms providers such as AT&T and BT. This access required a large amount of specialised and expensive equipment that had been preconfigured to function on the established telephone network.[33] The network access market has, though, changed and matured considerably in recent years. As recently as the late 1990s the market for internet access resembled the market that had first emerged in the 1970s. It remained quite immature, and the threat discussed by Professor Lessig was very real. At that time if I wanted to access the network I would still have to use dial-up access or turn to one of the emergent technologies for broadband access: either cable or ADSL/telephone access. It seemed that wires would carry access, and in Professor Lessig's words, controlling the wires meant controlling the network.

The market has changed substantially in the last five years: there is now plurality of access to the network. Now when I wish to access the network I have several access options and several access providers. At home I have a choice between traditional 'wired' access providers via an ADSL or cable connection to any one of dozens of ISPs, or I may use a wireless option provided by any one of several mobile telecommunications carriers. When I leave the house I have several wireless options, including, but not limited to a wireless card on my PDA, a wireless laptop and third-generation mobile phone handsets. My mobile phone accesses the network through my provider's network infrastructure, and next-generation devices such as Blackberry will even pick up my emails wherever I am. New laptops actively scan for available wireless access points and providers. I now have at least four access points to the network, when five years ago I only had one.[34]

This plurality of access will, I believe, increase over the next few years before there may be some natural (that is market-led) entropy. Cable and digital satellite companies will have a substantial role to play in network access through TV-on-demand and interactive services such as interactive news and sports. Traditional access providers such as telecoms companies

---

33 To buy a modem in the 1970s would cost in the region of $600. See Dwight, K, 'What About 56K Modems??', 1997, available at: *www.1960pcug.org/~pcnews/1997/december/what_about_56k_modems.htm*.

34 The four I use now are ADSL wired access, wireless access card, 3G mobile phone and digital television carrier.

and ISPs will continue also to play a key role, providing for high-speed access to the WWW, instant messenger, peer-to-peer and email services. These same companies will develop a synergy of wireless and wired access. Mobile telecommunications companies will have probably the biggest impact of the next five to 10 years. The delivery of music to MP3-enabled mobile phones through 3G connections is probably the biggest threat to the iTunes/iPod music model. Similarly, as the screen size/battery life problem is resolved, streaming video content will be the greatest challenge to traditional TV content providers – who doesn't want to watch the latest episode of their favourite programme while stuck on a train or in a traffic jam? The effect of this plurality of access is to demonstrate that the market in network access is now a mature and highly competitive market.

It is impossible to imagine that one company, or even a group of companies can take control of the access market and use this power to leverage control to the higher layers of the network, unless of course that is what the market wants. This means the decision as to what shape the network will take in the future is *our* decision – the decision of consumers – not access providers. To date, the evidence strongly suggests that although there are a large percentage of consumers who will trade freedom for convenience, or even fashion, there is a sufficiently healthy market in alternative systems and technologies to prevent control being centralised in the way suggested by Professor Lessig. The existence of such a mature market for network access suggests it is highly unlikely that the type of capture suggested by Professor Lessig will take place. Monopolies tend to arise in markets that are immature, such as Microsoft's control of the operating system (OS) market or AT&T's monopoly over the telephone market in the US, or where the associated costs of research and development (R&D) suggest economies of scale, as arguably is occurring in the pharmaceutical and petrochemical industries. The internet access market shows neither of these characteristics. Therefore it seems that should any company, or group of companies attempt to leverage control from the physical infrastructure layer to the higher layers, as described by professors Saltzer and Lessig, the market will route around this anomaly in the same way the network routes around damaged nodes. Thus if you want to use design tools to control cyberspace, you need to look higher – to the logical infrastructure layer.

Designing controls in the logical infrastructure layer

The logical infrastructure, or code, layer of the network, unlike the physical infrastructure layer, is not a mature market. In fact, due to the requirement that the various individual networks and protocols that make up the network are required to internetwork with each other, we require a single, or in competition terms, monopoly, carrier. Thus unlike the physical infrastructure layer where competition protects end-user consumers from unwarranted attempts to leverage access controls into wider network control, the logical infrastructure layer, due to its reliance on a single protocol and market immaturity is perfectly placed to leverage design-based regulatory controls from the logical infrastructure layer into the content layer; in fact this ability to so control lies at the heart of Joel Reidenberg's theory of *Lex Informatica*.[35] Reidenberg described *Lex Informatica* as a 'parallel rule system',[36] a system of technological architectures that can achieve similar regulatory settlements to that of legal regulation. In essence, what Reidenberg was describing here was the choice of alternative regulatory tools (legal and design tools) available to a hierarchical regulator. In examining the role of these design tools Reidenberg predicted several key applications of design regulation in cyberspace, including content blocking and filtering,[37] data privacy, data encryption and data management,[38] and digital rights management.[39] The development of commercial systems to deliver these requirements and their widespread application, both in cyberspace and the physical world, demonstrate that Reidenberg's theory was correct. But as to date, these systems have proven to be of limited scope and effectiveness and there is little reason to suspect that this will change in the future. For example the clear failure of technical protection measures (TPMs) such as Copy Control[40] and Cactus Data Shield (CDS)[41] on music CDs to prevent the proliferation of illegal MP3 music files on peer-to-peer file-sharing networks such as Kazaa, Gnutella and BearShare,[42] demonstrate the inability of such piecemeal design controls to gain effective control over even part of the network: in this case the distribution network for commercial audio files.[43] The reason such design-based solutions fail to implement complete control over the digital product or service they seek to protect is twofold. First, due to a layering

---

35 Reidenberg, J, 'Lex Informatica: The Formation of Information Policy Rules Through Technology', 1998, 76 Tex L Rev 553.

36 Ibid p 565.   37 Ibid pp 556–9.   38 Ibid pp 559–63.   39 Ibid pp 563–5.

40 *Copy control* is the name of a compact disc copy-protection system used on recent EMI CD releases in some regions. It has been the focus of intense consumer pressure with many claiming such discs do not work in all standard CD players.

41 CDS is owned and operated by Macrovision Ltd. It is the primary TPM used today. The latest version of CDS, CDS–300 v 7, claims to have eliminated compatibility problems.

42 As an experiment the author downloaded from one of the peer-to-peer providers the entire content of the new Garbage album 'Bleed Like Me' only three days after its worldwide launch and before it was available in the UK iTunes music store. The download was complete and perfect. It was subsequently deleted.

43 Similar failures of code-based design controls to effect comprehensive control over products or services may be seen in the DeCSS cases: *Universal City Studios Inc v Reimerdes*, 111 F Supp 2d 294 (SDNY 2000); *Universal City Studios Inc v Corley*, No. 00–9185 (2d Cir) and in relation to the failure of states such as China to implement leak-proof state firewalls against outside influences: see Deibert, R and Villeneuve, N, 'Firewalls and Power: An Overview of Global State Censorship of the Internet', in Klang, M and Murray, A (eds), *Human Rights in the Digital Age*, 2005, London: Glasshouse Press.

Network layer	Java MMS HTTP SMTP PDF VoIP	Content protocols
Carrier layer	TCP/IP	Carrier protocols

Figure 4.2 Network layers.

within the logical infrastructure layer it is impossible to leverage code-based controls across all network products or services, unless one can leverage control into the carrier layer (see Figure 4.2).

As can clearly be seen, content protocols, which may be found in the network layer, including proprietary protocols such as Microsoft's Microsoft Media Server (MMS) and Adobe's Portable Document Format (PDF) and open source protocols such as HTTP and SMTP, rely upon the TCP/IP carrier layer. Thus no network layer protocol is completely secure. Proprietary protocols, particularly those that seek to control the actions of end-users, are subject to attacks from crackers who seek to engineer around the controls that are imposed in the code.[44] Once these crackers successfully circumvent the security protocols of the software, the open access nature of TCP/IP means that there are no systems to prevent the distribution of illegally cracked software or content as the carrier layer does not, and cannot, differentiate between lawfully produced and illegal code. Thus control of specialised network code such as MMS cannot be leveraged into the wider network. This means that code-based controls, to be effective across the network, rather than simply within pockets of the network, must be implemented at the carrier layer. But herein lies the second reason why such code-based regulatory solutions are difficult, though not impossible, to implement in cyberspace. The carrier layer is founded upon a single carrier protocol – transmission control protocol/internet protocol (TCP/IP). TCP/IP was designed as an end-to-end protocol.[45] As it applies to the internet, the end-to-end structure dictates that putting intelligence in physical and link layers to handle error control, encryption, or flow control unnecessarily complicates the system. This is because these functions will usually need to be done at the end points anyway, so why duplicate the effort along the way? The result is an end-to-end network that provides minimal functionality on a hop-by-hop basis and maximal control between end-to-end communicating systems. This means though that there is no intelligence in the carrier protocols; any intelligence in the network

44 Crackers are hackers who break software security. They develop their own software that can circumnavigate or falsify the security measures that keep the application from being replicated on a PC.
45 Clark, D, 'The Design Philosophy of the DARPA Internet Protocols', 1988, *Computer Communications Review*, 18(4) p 106.

is built into the higher content protocols, which as we have seen can only be of piecemeal effect. To implement effective code- or design-based control systems across the network means therefore changing the current structure of the TCP/IP protocol in such a way as to introduce intelligence and the ability to control. It appears that Professor Reidenberg had anticipated some of these problems as long ago as 1998, as he noted in his *Lex Informatica* paper that:

> The shift in focus toward technical standards as a source of policy rules emphasizes technical fora whose institutions are not normally associated with governance. The Internet Engineering Task Force, the Internet Society, the World Wide Web Consortium, and traditional standards organizations like ISO, ETSI, and committees like T1 are the real political centers of Lex Informatica. Yet these groups are generally not governmental organizations. Rather, they tend to be consortia of interested persons and companies.[46]

The only bodies with the ability to leverage comprehensive code-based controls in the way described by Professor Lessig are therefore these technical 'consortia of interested persons and companies', but who are they?

The key thing to note about the technical standards bodies in question is that much like the TCP/IP protocol itself, they are of two distinct elements. As is clear from its name, TCP/IP is a two-part protocol. Transmission control protocol (TCP) is the network management protocol. It splits the content

---

46 See above, fn 35, pp 582–3. In turn the bodies he refers to are: (1) The Internet Engineering Task Force – a self-selected organisation that is the protocol engineering and development arm of the internet composed of network designers, operators, vendors and researchers; (2) The Internet Society is a non-governmental international organisation that seeks to co-ordinate internetworking technologies and applications for the internet; (3) The World Wide Web consortium is an international industry consortium run jointly by the MIT Laboratory for Computer Science in the United States and the Institut national de recherché en informatique et en automatique in France, which seeks to promote standards for the evolution of the Web and interoperability between WWW products; (4) The International Standards Organisation is a Geneva-based worldwide federation of national standards bodies from approximately 100 countries. Its objective is to promote the development of standardisation and related activities in the world with a view to facilitating the international exchange of goods and services, and to developing co-operation in the spheres of intellectual, scientific, technological and economic activity; (5) The European Telecommunications Standards Institute sets voluntary telecommunications standards for Europe and co-operates with the European Broadcasting Union for broadcasting and office information technology standards and (6) T1 (now ATIS – the Alliance for Telecommunications Industry Solutions) is a privately sponsored organisation accredited by the American National Standards Institute. It develops technical standards and reports regarding interconnection and interoperability of telecommunications networks at interfaces with end-user systems, carriers, information and enhanced-service providers, and customer premises equipment.

of a data file into smaller data packets making them easier to transmit. It then manages the transmission of these packets across the network ensuring that they are sent and received in the correct order and that network congestion and/or damage is routed around. TCP guarantees delivery of data and also guarantees that packets will be delivered in the same order in which they were sent. Internet Protocol (IP) is a less sophisticated, but no less essential, part of the network: it is the addressing protocol. It specifies the format of packets and the addressing scheme for the network. IP by itself is something like the postal system. It allows you to address a package and drop it in the system, but there is no direct link between you and the recipient. When combined with the management function of TCP though, to create TCP/IP, it allows the user to establish a connection between two hosts so that they can send messages back and forth for a period of time. Similarly, the technical standards bodies of the internet are split along these two functions: network standards and addressing standards; both are essential for the smooth functioning of the network, yet both, to a large extent, work independently. The key bodies dealing with these two elements are the Internet Society (network management) and the Internet Corporation for Assigned Names and Numbers (address management).

The Internet Society and network management

The Internet Society (ISOC) is the umbrella management organisation that co-ordinates and charters a wide variety of technical standards organisations who look after the day-to-day running and development of the network. ISOC's role is to provide guidance and leadership in designing and developing the future structure of the internet. The work of ISOC is organised around three themes, referred to as the 'ISOC pillars': (1) internet standards; (2) education; and (3) policy. ISOC co-ordinates, but does not directly manage, the development of new protocols and standards for the internet. It is the organisational home of key technical standards bodies including the Internet Engineering Task Force (IETF), the Internet Architecture Board (IAB), the Internet Engineering Steering Group (IESG) and the Internet Research Task Force (IRTF).[47] In this capacity, ISOC supports the request-for-comments editor (RFC editor) function of the IETF and issues internet standards advisories, publishes informational guides, and develops and conducts technical tutorials. In addition, ISOC acts as a channel to educate and promote standards internationally by providing training for IT professionals. To fulfil its educational remit, each year ISOC hosts two large open participation

---

47 All of these will be discussed below.

conferences[48] as well as organising an extensive series of smaller regional work-shops on subjects as diverse as Internationalised Domain Names (IDNs)[49] and Voice over IP (VoIP). ISOC is an open membership society which anyone may join. It offers individual membership at Member level (for payment of a $75 annual fee) and Associate Member level (at no cost). Both levels of membership allow for active participation in all ISOC activities in pursuit of the ISOC mission and goals,[50] and allow for attendance at, or receipt of, at all ISOC conferences, educational events, announcements, briefings and newsletters. Both member levels are able to participate in, and to have access to, ISOC's surveys and discussion groups, and both levels have the ability to participate in ISOC Chapters (see below). The only real distinction between both levels of individual membership is that those who join at Member level are able to participate in elections for ISOC board of trustee members.[51] In addition to the over 20,000 individual members ISOC has in more than 180 countries, there are over 100 organisational members. These are corpora-tions who support the work of ISOC and who donate between $2,500 and $100,000 annually to ISOC's work in this field. The internal management of ISOC is equally open. ISOC's executive management is carried out by a board of trustees. The board is made up of a maximum of 20 members who are elected in accordance with a rather complex selection system for a three-year term.[52] The board's main responsibilities are in managing the budget of ISOC and setting out the strategic plan of the organisation. Much of the work ISOC does to fulfil its three pillars is carried out by the 82 Local

---

48 The Internet Global Summit Conferences or INET, which is held annually at venues world-wide and Network and Distributed System Security Symposium Conference (NDSS) held annually in San Diego.

49 Internationalised Domain Names allow for a wider range of characters than the traditional domain name, which is restricted to the roman alphabet, causing problems in Arabic and Asian countries.

50 The ISOC mission statement is 'To assure the open development, evolution and use of the Internet for the benefit of all people throughout the world'. Its goals include: (1) development, maintenance, evolution, and dissemination of standards for the Internet and its internet-working technologies and applications; (2) growth and evolution of the internet architecture; (3) maintenance and evolution of effective administrative processes necessary for operation of the global internet and internets; (4) education and research related to the internet and internetworking; (5) harmonization of actions and activities at international levels to facilitate the development and availability of the internet; (6) collection and dissemination of information related to the internet and internetworking, including histories and archives; (7) assisting technologically developing countries, areas, and peoples in implementing and evolving their internet infrastructure and use; and (8) liaison with other organisations, gov-ernments and the general public for co-ordination, collaboration and education in effecting the above purposes.

51 Although there are quite complex rules on this which may be read at *www.isoc.org/isoc/general/trustees/select.shtml*.

52 The selection rules may be found at *www.isoc.org/isoc/general/trustees/select.shtml*.

Chapters. Reflecting the open and democratic nature of ISOC, anyone may form a chapter; all you need is 25 ISOC members and a set of local bylaws. It is usually Local Chapters that organise workshops and assist in hosting the INET conference. Local Chapters are also a strong voice in shaping policy and many chapter members also work for other standards-setting agencies like IETF or IAB. Thus we can see that ISOC is an open community: there are no barriers to entry and anyone who joins is asked to participate.[53]

At the technical and developmental level, the internet is made possible through creation, testing and implementation of internet standards, developed by a diverse group of standards bodies. ISOC's primary purpose is to host and provide funding support for these bodies. As stated above, ISOC is the organisational home to several such bodies. Primary among these is the Internet Engineering Task Force (IETF). The IETF is, like ISOC, a large open-access international forum. It is a community of network designers, vendors and researchers concerned with the evolution of internet architecture and the smooth operation of the internet. The actual development work of the IETF is done in its working groups. Most IETF members hold full-time jobs in the software, telecommunications or hardware sectors. They come together in their evenings and weekends to contribute to the work of the IETF. Because members of working groups may be found anywhere world-wide, much of the work is handled virtually via mailing lists. Working groups are collected into subject areas such as routing, applications, transport and security, and are managed by Area Directors. The Area Directors are in turn members of the Internet Engineering Steering Group (IESG). The IESG, as well as providing the Area Directors to co-ordinate the work of IETF working groups, also acts as the IETF's executive. It is responsible for technical management of IETF activities and the internet standards process. The IESG is directly responsible for developing new internet standards, including final approval of specifications as internet standards. IESG members are drawn from the IETF community, and in accordance with traditional models of internet management, IESG members are selected by the community because of their technical skills and service to the community.[54] Supervision and architectural oversight of the IESG and IETF is provided by yet another body, the Internet Architecture Board (IAB). The IAB is a committee of the IETF. It has particular responsibility for: (1) the confirmation of the IETF chair and IESG area directors, from nominations provided by the IETF; (2) oversight of, and occasional commentary on, aspects of the architecture

---

53 In terms of communications theory it is a 'deliberative democracy'. See further, Dryzek, J, *Deliberative Democracy and Beyond: Liberals, Critics, Contestations*, 2000, Oxford: OUP.

54 Among the technical community those with exceptional technical ability who provide service and guidance to the community are known as 'wizards'. The name seems to have developed from online gaming communities where wizards had powers, privileges and knowledge, which separated them from ordinary members.

for the protocols and procedures used by the internet; (3) oversight of the process used to create internet standards, including acting as an appeal board for complaints of improper execution of the standards process by acting as an appeal body in respect of IESG standards decisions; (4) editorial management and publication of the request for comments (RFC) document series; and (5) selecting the chair of the Internet Research Task Force (IRTF).

What is becoming clear very quickly is two related aspects of technical standards setting in cyberspace. The first is that there are a host of organisations that have overlapping functions, and who act as checks and balances on each other. No one body has authority to change an internet standard. Although the members of the IESG have the authority to adopt a new technical standard this decision is subject to appeal to the IAB and can only be taken once proper consultation has been made through the RFC process. Both the IESG and the IAB draw their authority from charters they have received from the Internet Society, which in theory could be withdrawn at any time. Thus any attempt to seize control of the IESG or the IAB by groups or individuals hoping to leverage design controls into internet standards appears to be futile. In addition, the IESG does not work in isolation. Although IESG members, in their role as area directors, chair IETF working groups, membership of working groups is open to all, as is membership of the IRTF, the sister organisation of the IETF, which provides long-term planning for the network. When you attempt to map the relationships between these bodies, as I have attempted to do in Figure 4.3, a complex web of oversight, management, delegation and authority begins to emerge.

These relationships reflect the network itself and the men (and it is mostly men) who participate in these fora. The reason the relationships between committees such as the IAB, the IETF, the IESG and the IRTF seems to be so ad hoc is because that is exactly what they are. Each of these committees grew up because there was a particular problem with the management and development of the network protocols which needed to be dealt with, and for which at the time no one had specific responsibility. Someone – usually one of the early network pioneers such as Vint Cerf, Jon Postel or Larry Roberts – would take over management of the problem, but in doing so they would seek the input of their peers. To do this a mailing list would be set up and management of the problem would be delegated to the members of that particular task force or group. Over the last 10 years the network has grown beyond imagination, but the ethos of management by consensus still applies. This is the second key aspect of these organisations. The ethos of bottom-up consensus building was at the heart of the development of the network from its inception in the early 1960s until the creation of TCP in 1974. Today it remains at the heart of all technical developments and decision making in cyberspace: made possible by the participation of thousands of people from throughout the world by means of the open participation mailing lists of the IETF and the IRTF. The management of TCP/IP and other internet

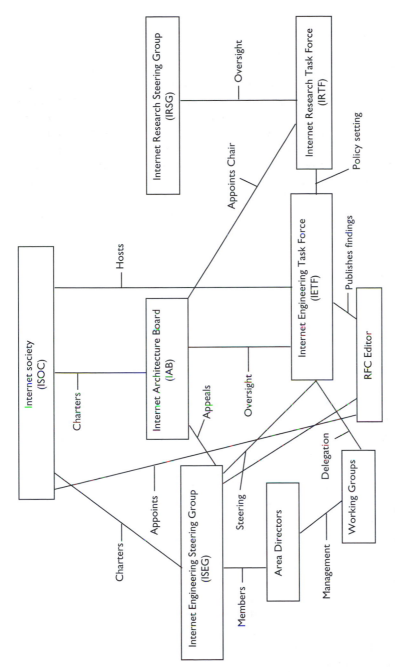

Figure 4.3 Structure of network regulators.

protocols is therefore as open and inclusive as the software protocols themselves. There is no opportunity for agency capture by external bodies. Any
attempt to do so would be detected by the members of the decision-making
community and would be defeated by the labyrinthine process of checks and
balances found in the network regulatory structure. It may be difficult to
explain how all these bodies are related, but what is clear is that the checks
and balances in place protect the open nature of the network, ensuring an
open-source platform for the future.

ICANN, IANA and location

The story of TCP and network protocol management is, though, only half
of the story of regulation at the logical infrastructure layer. Whereas the
management of the TCP/IP protocol by the Internet Society and related
technical standards organisations, regulates how, and what, may be carried on
the network, this is not the only factor that decides whether your content is
accessible on the internet. The TCP protocol may allow you to post content
to the network, but if your content cannot be found then posting the content
is of no value. The internet is a vast network of communications networks.
Like all communications networks it relies upon the ability to deliver content
to another party. The telephone network functions because each telephone
(or node) connected to the network has a unique identifier. Thus to telephone
an individual node in London from New York, one follows a hierarchy of
numerical identifiers, which allows the exchanges on the network to route the
call to the correct recipient node. First, you select the country by dialling the
correct country code for the UK (44); this is followed by a regional code,
which for London is 20; next, you dial an exchange code; in the case of the
London School of Economics, this is usually 7955; then the actual code for
a single telephone or node on this exchange, for example the law department
fax number is 7366. Thus that individual node on the telecommunications
network may be called from New York (or anywhere else) by dialling in the
descending identifier, 442079557366. This is the unique identifier of one node
on the network and by using this communication may be established between
any two network nodes. The mail network works in a similar fashion. Unlike
the telephone network it uses an ascending identifier, starting with the narrow
geographical location of a person (or building or room) and building up to a
national identifier. Thus to send a postal message to me at my office one opens
with my office number: *A372*; then builds outwards to the building location
of this office – *Old Building*; then street name – *Houghton Street*; City –
London; and finally country – *United Kingdom*. Systems of postcoding have
been developed to make the system more efficient by assisting the postal
service in large cities such as London, where street names are often repeated
in different areas of the city. Thus the postcode for the LSE is *WC2A 2AE*
and a letter sent from New York simply addressed to Andrew Murray, Room

A372, Old Building, WC2A 2AE, United Kingdom should reach me with little difficulty. What these two examples demonstrate is that the key component of any communications network is its addressing protocol. Without the number of another telephone to call, a telephone is simply a worthless collection of plastic and wires, while without addresses letters become merely scrap paper. The internet, quite obviously, has its own addressing system built on the IP component of TCP/IP. Clearly, whoever controls this wields considerable power in cyberspace as they could excommunicate whole sections of the network should they wish to do so. So how do you address internet communications and who manages the system?

As internet communications are communications between computers the need for a linguistic addressing tool seemed minimal when the addressing protocol for the internet was developed. The primary addressing tool is therefore a numerical identifier called an internet protocol, or IP, address. As with a telephone number or a postal address this number must be unique to allow for the smooth flow of information within the network. IP addresses are used when browsing the web to enable the transmission of communications between the user's web browser and the server hosting the website. They are also used in the header of email messages and, in fact, are required for all programs that use the TCP/IP protocol. Version 4 of the internet protocol (IPv4), which is the current standard IP protocol, uses an IP address consisting of 32 bits,[55] usually shown as a 'dotted quad': four octets of binary numbers represented in decimal form in the range 0–255.[56] For example the IP address of the computer I am currently using is 158.143.112.199, but as computers do not work in decimal notation this will be converted by the network servers and routers into binary notation and be read by them as 10011110.10001111.1110000.11000111. As it is easier for humans to remember decimals than it is to remember binary numbers, we use decimal notation to represent IP addresses when describing them. The address space of the IPv4 protocol (the number of available unique identifiers allowed) is 2^{32} or 4,294,967,296 unique host interface addresses. At the time it was adopted this seemed to be an almost limitless supply of addresses,[57] but as the network has developed the strain on this resource has become quite

---

55 A bit, short for *binary digit*, is the smallest unit of information on a machine. A single bit can hold only one of two values: 0 or 1. More meaningful information is obtained by combining consecutive bits into larger units. For example a byte is composed of eight consecutive bits.

56 The maximum value of a single octet under the currently most widely used IP protocol IPv4 is 255 as in binary notation this is 11111111; to record 256 would require a ninth binary figure, not available in a digital octet.

57 IPv4 was developed in the 1970s and adopted in September 1981 when RFC 791 set out the new Internet Standard TCP/IPv4 suite (RFC 791 available at: *www.ietf.org/rfc/rfc791.txt*). At this time there were only 213 internet hosts (see Lottor, M, *Internet Growth (1981–1991)*, 1992, RFC 1296, available at: *www.ietf.org/rfc/rfc1296.txt*) and so the prospect of using over 4 billion addresses seemed impossible.

heavy.[58] For this reason a new version of the IP protocol, IPv6,[59] has been developed by the Network Working Group of the IETF.[60] In IPv6, addresses are 128 bits rather than 32 bits. This will allow 2^{128}, or about 3.403×10^{38}, unique host interface addresses: a mind-boggling availability of unique identifiers.[61] An IPv6 address is written as eight four-digit hexadecimal numbers separated by colons. Thus to the human eye it looks something like: 3FFE:FFFF:0100:F101:0210:A4FF:FDE3:9566.[62]   Computers, including hosts and routers will continue to read this as binary notation, but space prevents me from reproducing the binary equivalent of a hexadecimal IPv6 address. Both IPv4 and IPv6 addresses are descending unique identifiers like telephone numbers. You read them from left to right with the first two sets of quads being used to identify the network location of your computer. For example all computers on the LSE network are allocated an IP address, which begins 158.143, while King's College, London operates from the 137.73 address space, while University of Oxford IP addresses all begin 163.1.[63] Who, though, decided this? How are these addresses allocated? While the actual IP protocols – IPv4 and Ipv6 – were developed in the now relatively familiar working groups of the IETF, there is nothing in the protocol that determines the allocation of IP addresses: to do this there are a whole group of other regulatory bodies. These are the IP Registries, bodies who deal with the allocation and management of the IP addressing system.

Historically the role for allocating IP addresses, and the now more important mnemonic identifier of an IP address, its associated domain name (of which more below), was that of one man. The man was Jon Postel. We have already heard how Postel was intimately involved in the development of the TCP/IP protocol, but designing protocols was not the major part of his work. For nearly 30 years Jon Postel was probably the most powerful of the network 'wizards', who modelled, managed and developed the

58 There are now 394,991,609 host computers, still some way short of the 4 billion plus supported by IPv4, but with the arrival of many more network devices, in particular WAP and 3G mobile phones, wireless office devices such as Blackberries and the increase in 'always on' broadband access in place of dial-up access, which allowed for sharing of IP addresses, some believe that IPv4 addresses may start to run out in the next 10 years.

59 IPv5 was an experimental (non-addressing) protocol for the real-time streaming of content, which was not developed. Hence the leap from IPv4 to IPv6.

60 Hinden, R and Deering, S, *IP Version 6 Addressing Architecture*, 1998, RFC 2373, available at: *www.ietf.org/rfc/rfc2373.txt*.

61 In unabridged notation this is: 340,300,000,000,000,000,000,000,000,000,000,000,000 unique addresses or taking the world's population to be in the region of 6 billion people, 56,716,666,700,000,000,000,000,000,000 devices per person.

62 In hexadecimal notation A=10; B=11; C=12; D=13; E=14 and F=15.

63 Thus it should be obvious that IP addresses are not specific to individual nations, but follow a different allocation system.

network.[64] He was the RFC editor from the date of RFC-1, which was published on 7 April 1969, until his death on 16 October 1998: a role that helped shape the network through its formative stages in the 1970s. Despite being a key network position this was not the major part of Postel's working life. For nearly 30 years Jon Postel personally allocated blocks of IP addresses to those who needed them. Originally, as with nearly all other network management roles, Postel took the job on out of necessity. In 1969 Jon Postel joined the Network Management Centre at UCLA; the centre was playing a major role in building the ARPANET and one of the problems was that addressing the protocols that had been developed to allow for the flow of information between nodes needed to be centrally managed and allocated to prevent errors or repetition. As was later recorded by Vint Cerf: 'Someone had to keep track of all the protocols, the identifiers, networks and addresses and ultimately the names of all the things in the networked universe.'[65] The person who quietly and efficiently took on this role was Jon Postel, who first started keeping lists of network protocol numbers on a scrap of notebook paper. From the late 1960s to the early 1980s Jon Postel, through his office, first at Network Management Centre, then later at his office in the Information Sciences Institute (ISI) at the University of Southern California, allocated new IP address blocks and mnemonic identifiers to new network nodes. Initially this was a straightforward task. In its first year, 1969, only four nodes were added to the network. By 1971 the network had expanded to only 23 hosts with the network not breaking the 200 host barrier until 1981, when the number of hosts reached 213. This meant on average between 1969 and 1981 Postel needed only to allocate IP addresses to around 18 additional hosts per annum. In the early 1980s though, the speed of development of the network rapidly increased. Between May 1982 and August 1983, 327 new network hosts were added, and between September 1983 and October 1984, a further 462 new hosts were added. Postel could no longer keep up with requests for new IP addresses: it was taking up all his time and his other work with the ISI was being affected. Postel decided he needed help in managing this task so he approached the US Department of Defense (USDoD) for funding and in 1983 set up, with his new assistant Joyce Reynolds who also worked at the ISI, a new organisation to take over the newly onerous task of allocating IP blocks. Postel called his new organisation the Internet Assigned Numbers Authority (IANA), but IANA really existed, and to most extent still exists, only as a name. To all intents and purposes IANA does not exist.

---

64 In *The Economist*, Vol. 342 Iss 8003 (8 February 1997), it was noted that: 'God, at least in the West, is often represented as a man with a flowing beard and sandals. Users of the internet might be forgiven for feeling that nature is imitating art – for if the Net does have a god, he is probably Jon Postel, a man who matches that description to a T.' (p 88).

65 Cerf, V, *I Remember IANA*, 1998, RFC 2468, available at: *www.ietf.org/rfc/rfc2468.txt*.

It has no charter,[66] no bank accounts and no actual staff. Rather, funding given to IANA from the USDoD and other sources[67] was paid through the accounts of the ISI and was used to pay for around four full-time staff of the ISI to carry out the IANA function. Despite this, IANA remains one of the key regulatory bodies of the IP addressing system. IANA is still responsible for the allocation of new IP addresses, although nowadays with millions of new nodes being added to the network annually it is impossible for IANA with its meagre budget and small staff to do this alone. Management of the IANA function is now decentralised through an extensive support network, which takes on much of the everyday work of actually allocating IP addresses. Primary among these supporting networks are the five Regional Internet Registries (RIRs), who manage IP addresses within their geographical region. They ask IANA to allocate large blocks of IP addresses to them, then the five: the African Internet Numbers Registry (AfriNIC); the Asia Pacific Network Information Centre (APNIC); the American Registry for Internet Numbers (ARIN); the Latin American and Caribbean Internet Addresses Agency (LACNIC) and the Reséaux IP Européens Network Coordination Centre (RIPE NCC), pass on blocks of available IP addresses to Local or National Internet Registries such as Nominet UK, who in turn pass on IP addresses to ISPs and other network access providers for allocation to individual nodes (see Figure 4.4, below). Thus the key task that Jon Postel took on in 1969 – that of allocating network addresses to new nodes – is still undertaken by IANA. It should though be noted that IANA has no scope for regulatory intervention or management. The availability of IP addresses is controlled by the IP protocol. IANA cannot create or delete IPv4 addresses, only make them available for use. In fact it is this rigidity of the IP protocol that drove the development of IPv6. Equally, RIRs and Local Internet Registries can do no more than allocate numbers when requested. These bodies are administrative rather than regulatory. This is not to say that these bodies are not worthy of study by students of cyber-regulatory theory, because in addition to the basic IP addressing function discussed, they also play a role in a much more important social function: maintenance and development of the Domain Name System (DNS).

As we have seen, IP addresses, although the ideal tool for computers, are quite user-unfriendly from a human point of view. A binary IP address such as 10011110.10001111.1110000.11000111 is completely meaningless and therefore totally unmemorable. Even its decimal equivalent of 158.143.112.199

---

66 Although an IANA Charter is listed as a 'work-in-progress' in RFC 2028: Hovey, R and Bradner, S, *Best Current Practice*, 1996, RFC 2028 (available at: *www.ietf.org/rfc/rfc2028.txt*), it has to the best of the author's knowledge never been completed and/or published.

67 Postel has registered the .us country code top-level domain in his name and by 1996 was making in excess of $200,000 per annum for IANA by licensing registrar Network Solutions Inc. to manage the domain.

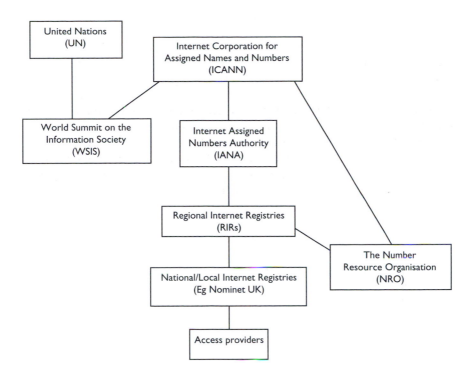

Figure 4.4 Structure of address protocol regulators.

is considerably more complex than a telephone number as there is a lack of an underlying geographical structure in the IP addressing system. Both the London School of Economics and Kings College London share the first five digits of their international telephone number (44207) as both are geographically close, but their IP addresses are quite dissimilar (158.143 and 137.73). This lack of geographical indicator means users would need to remember a number of between 9 and 12 digits, somewhat more than the seven-digit number most average people can recall.[68] When ARPANET was created this was not a problem; only a few addresses needed to be remembered and at that time addresses were shorter as the IPv4 protocol had yet to be developed. By the early 1970s though, more hosts were being added and Peggy Karp, one of

---

68 This figure has been arrived at following much research into the physiology and psychology of memory. See eg Baddeley, A *et al*, 'Random Generation and the Executive Control of Working Memory', 1998, 51(4) *Quarterly Journal of Experimental Psychology*, 819 Brahms, S, *The US Phone Nomenclature System as Applied to Dynamic Knowledge Repositories: A Recap of Bell Labs Research*, 2001 Hasting Research, available at: *http://www. hastingsresearch.com/ net/05-nomenclature.shtml*. This may (at least in part) explain why we find it more difficult to remember mobile phone numbers.

the early authors and editors of the RFC series proposed the introduction of 'host mneumonics' to overcome this problem.[69] Her proposal was for the creation of a text file that mapped the addresses of all network resources. This file, called *Hosts.txt*, could be installed on a local server, then when the user wished to access a remote server the local server would look up the network address of the remote server based upon a mnemonic name entered by the user. Like Jon Postel's management of network addresses, management of this new mnemonic address list was taken on in an ad hoc fashion. Whenever a new machine was added to the network all its details would need to be manually added to the *Hosts.txt* file by the staff of the ARPANET Network Information Centre at the Stanford Research Institute.[70] The problem with a manual system though is that as the network grew, the *Hosts.txt* file grew leading to an increasing amount of updates and the occasional error in the system and as these problems grew they eventually led to network engineers conceiving of a replacement for Karp's text-based mnemonic addressing system. The concept of the DNS was first conceived in RFC 799, *Internet Name Domains*, written by David Mills of COMSAT and published in September 1981.[71] Mills saw his system as being scalable, something that with its manual updates the *Hosts.txt* file was not. He imagined that an internet DNS could facilitate, 'thousands of hosts' rather than the 400 or so on the *Hosts.txt* file at that time. As we now know he undersold his idea spectacularly, there being 394,991,609 active hosts on the DNS in January 2006.[72] Mills' idea was picked up by Jon Postel, who with Zaw-Sing Su from the Stanford Research Institute, outlined the structure of the DNS and how it would allow for access in RFC 819.[73] The system uses a tree model with a 'root' at the bottom, anchoring the rest of the DNS and branches building upward and outward. At the base are the 'root servers', which are the foundation of the DNS. The root, theoretically, knows where to find the names and IP addresses of all network hosts. The root server network actually consists of 13 servers (the maximum allowed under IPv4) spread around the world. The reason we have a network of servers rather than a single server is twofold: first, to have one server handle this task would put it under enormous strain, and second, having just one root server would leave the network open to attack at that

69 Karp, P, *Standardisation of Host Mneumonics*, 1971, RFC 226, available at: *www.ietf.org/rfc/rfc226.txt.*
70 The operator of the new host would be required to fill out an email template and send it to the team at the ARPANET NIC, who would then update the *Hosts.txt* file. Operators of network hosts would receive new versions of the *Hosts.txt* file with these updates at regular intervals for installation on their host.
71 Mills, D, *Internet Name Domains*, 1981, RFC 799, available at: *www.ietf.org/rfc/rfc799.txt.*
72 Source: Internet Systems Consortium Domain Survey: Number of Internet Hosts, available at: *www.isc.org/index.pl?/ops/ds/.*
73 Su, Z and Postel, J, *The Domain Naming Convention for Internet User Applications*, 1982, RFC 819, available at: *www.ietf.org/rfc/rfc819.txt.*

point. Of the 13 servers though there is one of special significance: the 'A' root server, based in Dulles Virginia contains what is deemed to be the definitive DNS database. Thus the other 12 root servers regularly check their database with the 'A' server and make changes based upon the records of the 'A' server. This is to prevent any discrepancy in the databases of the servers that make up the root server network. Thus the 'A' server holds the definitive DNS database, and whoever controls the 'A' server is in a particularly strong position within the DNS system.

The 'A' server is updated by constantly gathering in information from its offspring servers, which are maintained by the RIR. They in turn obtain their updates from Local or National Internet Registries, and so on down the line. When you log on to the network via your ISP and enter a domain name your ISP then uses the DNS to match up the mnemonic domain name you have entered with the underlying IP address that you are seeking. Thus assuming I want to read a news story on the *Wired* website, I would enter the domain name *www.wired.com* into my web browser. To get the correct IP address for this page my ISP must carry out a DNS search. Its first port of call will be to check its own name server. If you have a large ISP like Orange or AOL, and the page you are visiting is popular it is likely that another of its customers will have visited this page recently and it will have a stored, or cached, copy of the DNS details of the page you seek. If, though, your ISP does not have the details, it will forward this on to one of the root servers. The root server knows the IP addresses of the name server that manages .com top-level domain (TLD).[74] It gives this IP address to the name server of your ISP, which then sends a query to that name server asking for the IP address of *wired.com*. The name server for the .com TLD will know the IP addresses for the name server handling the *wired.com* domain, so it returns this IP address to your ISP's name server. Your ISP's name server then contacts the name server for *wired.com* and asks it for the IP address for *www.wired.com*. Your browser can then contact this server and request the web page (see Figure 4.5). This of course all happens in the blink of an eye and the user is unaware of all these transactions taking place. At the heart of this system though is the name server database managed by the root servers. Without the root server, each individual name server would need to be continually informed of all changes within the name server network. The root servers, as their name suggests, anchor the entire DNS, and therefore those parties that control the root servers are in a particularly powerful regulatory position.

In 1983, the Defense Communications Agency of the US Department of Defense (DoD) chose the Stanford Research Institute (SRI) to manage the new DNS and to operate the root server. The new DNS began operations on 15 March 1985 with the registration of *symbolics.com*, the first .com domain

---

74 The structure of domain names will be discussed further below.

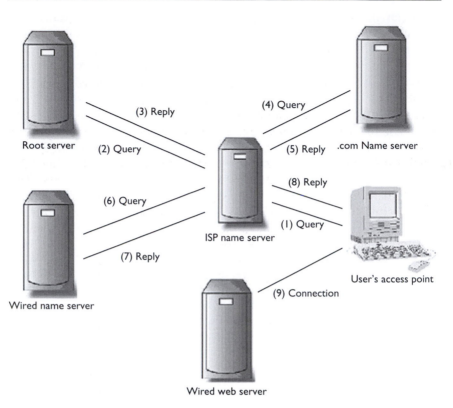

Figure 4.5 The DNS system.

name.[75] The SRI managed the DNS and the root throughout its early devel-
opmental phase in the 1980s, until in 1991 the Defense Communications
Agency (DCA), now known as the Defense Information Systems Agency
(DISA), put management of the DNS and root out to competitive tender.
The winning bid came from a defence contractor known as Government
Systems Inc., which in turn outsourced the work to a small private contractor,
Network Solutions Inc. (NSI). For the first time, management of the DNS
and root was in private hands. Network Solutions sought to reinforce their
position by bidding for further management tenders with the US National
Science Federation including tenders to manage the registration of domain
names and IP number assignment for the NSFnet, the major civilian internet
backbone at the time. By 1992, NSI controlled the DNS and root, was the
registrar for domain names on the network and was responsible for allocating
new IP addresses, which they were allocated in bulk from IANA. Control in

75 *www.whoisd.com/oldestcom.php.*

the DNS had, to a large extent, been centralised in NSI. Then, in 1993, when CERN gave notification that they were putting the HTML code into the public domain, the value of the network, and NSI's contracts, skyrocketed. Domain names at this point were being given away for free, NSI's profits came from its contracts with the DoD and the NSF. But with the number of domain name registrations, in particular speculative and/or abusive registrations such as business.com and mcdonalds.com,[76] rising rapidly, NSI sought permission to charge for domain name registrations and in 1995 they began charging a $50 annual fee per domain name registered. Interestingly, the actual cost for processing and managing registrations was in fact around $35. NSI were not, though, profiting from their domain name business. The excess funds generated were put into a separate fund, the Internet Intellectual Infrastructure Fund, which had been created to offset government funding for the preservation and enhancement of the internet. Between 14 September 1995, when the $50 fee was introduced,[77] and 1 April 1998 when the Internet Intellectual Infrastructure Fund portion of the registration fee was removed, this fund generated more than $45.5 million for the US Treasury. The collection of these funds was stopped after a successful legal challenge in the case of *William Thomas et al v Network Solutions Inc and National Science Foundation*,[78] which challenged the Internet Intellectual Infrastructure Fund portion of the fee on the basis that it was an illegal taxation that had been raised without the agreement of Congress.[79] Nevertheless, simply by charging a registration fee, NSI sparked off a period of great dispute pertaining to the management of the DNS. The internet 'old guard', led by Jon Postel were outraged that a private monopoly had been created in the DNS, and that that monopoly could charge for what they saw as a basic service. Postel proposed that NSI's monopoly be broken up, that IANA be given formal

---

76 Business.com was speculatively registered and sold twice, once in 1997 for $150,000, then again in 1999 for $7,500,000. Mcdonalds.com was 'abusively' registered in breach of the McDonalds Trade Mark by *Wired* Journalist Joshua Quittner in 1994 to illustrate the problem of 'Cybersquatting'. The full story is at: *www.wired.com/wired/archive/2.10/mcdonalds.html*.

77 'Amendment 4 to Cooperative Agreement Between NSI and U.S. Government', 13 September 1995, available at: *www.icann.org/nsi/coopagmt-amend4–13sep95.htm*.

78 Civ. No. 97–2412 (TFH).

79 The Judge, Judge Hogan, noted: 'The Preservation Assessment is clearly a tax . . . There is no dispute that the assessment is involuntary – it is automatically charged to every domain registration, and registrants cannot opt out of the charge. Further, NSI collects the assessment for the government's use on public goals, and not in any way to defray regulatory cost. In addition, defendants admit that the cost of providing the initial registration service is at most $70; the $30 paid to the Preservation Assessment is entirely in excess of cost. Thus, there is no dispute that the Preservation Assessment exists to generate revenue for public projects and goals, or that it is a fee imposed independent of and above the cost of domain name registration. For these reasons, the Preservation Assessment is not a regulatory fee, but is instead a tax on registration.'

legal foundations and that the Internet Society take over regulation of the domain name registration process.[80] To give effect to Postel's proposal ISOC created a working group to take it forward. The group known as the International Ad-Hoc Committee (IAHC) announced they would create seven new top-level domain names that would compete with .com and NSI's monopoly.[81] The proposals of the International Ad-Hoc Committee, now snappily entitled, *A Memorandum of Understanding on the Generic Top Level Domain Name Space of the Internet Domain Name System* or *gTLD-MoU*,[82] polarised debate. Many American commentators were concerned that these proposals meant that the Swiss-based, World Intellectual Property Organisation (WIPO) and International Telecommunications Union (ITU) would take on much of the management of the DNS. Others, fearing it would lead to a conflict between NSI and ISOC, attempted to model alternatives to the Committee's proposal. One alternative mooted was the Enhanced Domain Naming System (eDNS). This proposed a substitute root server system outside the traditional root, which would be used to administer top-level domain names not already in use. This proposal was simply an extension of a common alternative DNS system known as Alternate Roots or Alt. Roots. The problem with setting up Alt. Roots is that while they are technically easy to set up, actually running a reliable root server network in the long term is a serious undertaking. It requires multiple servers to be kept running 24 hours a day, seven days a week in geographically diverse locations. During the dot.com boom of the late 1990s, several such Alt. Roots were set up, but only a few remain, the largest of which is OpenNIC, which offers alternate TLDs such as .parody, .indy and .geek.[83] The problem with Alt. Roots is that only a small proportion of ISPs actually refer to any of the Alt. Root servers. This means the user cannot access any of the sites served by Alt. Root operators unless they specifically tailor their web browser to look for these servers and to do this they need to have the IP address of the root server in question, clearly not a job for the 'casual surfer'. Thus the eDNS proposal, like most Alt. Roots, did not survive long and the debate quickly returned to who should manage the established root DNS. The White House of President Clinton had, by his time, entered the debate directing that the DNS should be privatised and that competition within the DNS should be increased.[84] Then on 17 July 1997, human error at NSI caused the master files for the .org, .net and .com

80 Postel, J, *New Registries and the Delegation of International Top Level Domains*, 1996, available at: *www.watersprings.org/pub/id/draft-postel-iana-itld-admin–01.txt.*

81 *Final Report of the International Ad Hoc Committee: Recommendations for Administration and Management of gTLDs*, 4 February 1997, available at: *www.iahc.org/draft-iahc-recommend–00.html.*

82 Available at: *www.iahc.org/gTLD-MoU.html.*   83 *www.opennic.unrated.net/tlds.html.*

84 The White House, *A Framework for Global Electronic Commerce*, 1 July 1997, available from: *www.technology.gov/digeconomy/framewrk.htm.*

TLDs to become corrupted. Although the problem was quickly repaired there were knock-on effects throughout the following day. All parties that had previously not recognised the full importance of the root DNS were given a lesson as to the instability of the network: if NSI pulled the plug the network would become unstable almost instantly.

With the events of 17/18 July as a backdrop, the White House was spurred on to further action. Ira Magaziner, a senior policy advisor to President Clinton, was put in charge of the process. He was keen to ensure that the system was privatised, but that equally there was no chance of the system failing once it was in private hands. He took much of the framework of the IAHC's gTLD-MoU, but with one key difference: the US Government, not the Internet Society, would oversee the transition of management from NSI to the new non-profit organisation that would be set up to take on the DNS management role of NSI.[85] The 'old guard', being predominantly academics, did not take well to what they saw as unwanted government interference and on 4 February 1998, Jon Postel carried out a 'transition plan test' to, in his words, 'test how easily management of the root servers could be passed to another machine'.[86] He reconfigured the root server system so that the other servers got their DNS information not from the 'A' server operated by NSI, but rather from the 'B' server at the Information Sciences Institute, which he controlled. The White House reacted coolly to Postel's test, calling it 'unfortunate'. Although the White House did not seem to grasp the significance of Postel's experiment, or perhaps had chosen to play it down, Jon Postel had shown the weakness in the system: the DNS can only work with the agreement of all involved. Although NSI were in the position of controlling the 'A' server, and although NSI at that time had a monopoly over the registration of the generic TLDs (gTLDs)[87] including .com, this could all be removed by a simple line of code. The network, although appearing robust to most users has some points of fragility. Previously the benevolence of the network pioneers, who were mostly academics and researchers, had ensured that there were no conflicts or competition to control valuable network resources. Now with the commercial value of the network increasing on a daily basis, competition for control over valuable network resources such as the DNS root had led to a bitter dispute between the old guard

85 The White House, *A Proposal to Improve Technical Management of Internet Names and Addresses*, 30 January 1998, available from: *www.ntia.doc.gov/ntiahome/domainname/ dnsdrft.htm*.

86 Gittlen, S, 'Taking the wrong root?', *Network World*, 4 February 1998, available at: *www.nwfusion.com/news/0204postel.html*.

87 Generic top-level domains or gTLDs are domain names of an international character open to any interested party. In 1998 they were .com, .org and .net and may be distinguished from country code top-level domains such as .uk and specific top-level domains such as .edu reserved for educational establishments.

and new commercial interests. In an attempt to preserve the values of the network pioneers Jon Postel began working alongside the US National Telecommunications and Information Administration, the body given management of the White House DNS project. He drew up the bylaws for what he called 'the new IANA', set out that the body should have responsibility for IP addresses, domain names and protocol parameters, and designed the new body to have responsibility for the root server. On 1 October 1998, the new regulatory and management body for internet addressing was named: it would be known as the Internet Corporation for Assigned Names and Numbers (ICANN). The new regulator would be a California-based not-for-profit entity managed by a representative board of directors drawn from around the world.[88] By the terms of its Articles of Association it was authorised to take responsibility for:

> the operational stability of the Internet by (i) coordinating the assignment of Internet technical parameters as needed to maintain universal connectivity on the Internet; (ii) performing and overseeing functions related to the coordination of the Internet Protocol (IP) address space; (iii) performing and overseeing functions related to the coordination of the Internet domain name system (DNS), including the development of policies for determining the circumstances under which new top-level domains are added to the DNS root system; (iv) overseeing operation of the authoritative Internet DNS root server system; and (v) engaging in any other related lawful activity in furtherance of items (i) through (iv).

On 25 November 1998 in a memorandum of understanding (MOU) executed between the US Department of Commerce (DOC) and ICANN, ICANN was formally awarded management of the root server, DNS and the IANA function.[89] By that time Jon Postel was dead.[90] The one man who truly

---

88 By Section 6 of the Original Bylaws as passed on 6 November 1998, 'In order to ensure broad international representation on the Board, no more than one-half of the total number of At Large Directors serving at any given time shall be residents of any one Geographic Region, and no more than two of the Directors nominated by each Supporting Organization shall be residents of any one Geographic Region. As used herein, each of the following shall be a "Geographic Region": Europe; Asia/Australia/Pacific; Latin America/Caribbean Islands; Africa; North America. The specific countries included in each Geographic Region shall be determined by the Board, and this Section shall be reviewed by the Board from time to time (but at least every three years) to determine whether any change is appropriate.' See *www.icann.org/general/archive-bylaws/bylaws–06nov98.htm.*
89 Memorandum of Understanding Between the US Department of Commerce and Internet Corporation for Assigned Names and Numbers, 25 November 1998, available at *www.icann.org/general/icann-mou–25nov98.htm.*
90 Jon Postel died of complications following heart surgery on 16 October 1998.

understood the IANA function, the root server system and the DNS had been snatched away just when he was most needed.

ICANN fulfils a similar role to that of the Internet Society, but while ISOC evolved naturally from within the engineering network to provide a framework for development and evolution of the network protocols, ICANN was created artificially by external forces to remove a monopoly barrier and to centralise co-ordination of several addressing tasks that had become dispersed over time. One of the first key tasks ICANN faced was the development of policies regulating the management and allocation of domain names to allow for competition in the registry market for the generic gTLDs, .net, .org and .com. To this end, ICANN set to work on its first active project – creating a register accreditation system that would allow new registries to enter the market. Almost immediately though they met resistance. Trade mark holders were understandably apprehensive about any changes to be introduced into the market for domain names. They were concerned that by creating competition in the market for gTLDs alternative routes for cybersquatting, the practice of unlawfully registering the name or mark of a well-known individual or entity as a domain name with a view to later selling that name at a profit, would open up. As a result of their fears, trade mark holders sought specific assurances that the deregulation of the gTLD market would not allow for the creation of alternative gTLDs[91] without systems having first been put in place to assure their marks adequate protection within these new markets.[92] Resistance also came from NSI, which at this time had the monopoly right to manage the 'A' Root Server and through this the registration system for gTLDs. The Department of Commerce had anticipated this problem and had negotiated an agreement which required NSI to provide registrars approved under the new ICANN accreditation shared system access to its registry services in return for payment of a flat fee agreed by NSI and the Department of Commerce.[93] The DOC and ICANN announced that following this agreement initial experiments with registrar accreditation were a success: from 21 April 1999 five new registrars were to be allowed access to the lucrative .com, .net and .org gTLDs through the shared registry system

91 Alternative gTLDs such as .biz, .info and .name had long been proposed as a simple way to create new markets and therefore competition. It was envisaged by many that Network Solutions could retain their monopoly in .com while alternate domains such as .biz were promoted by competing registries. Unfortunately the market penetration of the .com brand meant this was unlikely to succeed.

92 World Intellectual Property Organisation, *The Management of Internet Names and Addresses: Intellectual Property Issues: Final Report of the WIPO Internet Domain Name Process*, 30 April 1999, ch 5, available at: *http://arbiter.wipo.int/processes/process1/report/pdf/report.pdf*; US Department of Commerce, *Management of Internet Names and Addresses*, Docket Number: 980212036–8146–02, 5 June 1998 at para 8, available at: *www.icann.org/general/white-paper–05jun98.htm*.

93 Mueller, M, *Ruling the Root*, 2002, Cambridge, MA: MIT Press, pp 187–8.

that the DOC had negotiated with NSI.[94] In truth, though, NSI were continuing to flex their considerable muscle outside of ICANN's control. By the middle of 1999 NSI were still not accredited by ICANN, yet remained the largest provider of .com registrations. Worse still, under NSI's contract with the DOC it could subcontract access to the NSI registry to agents who were not recognised or accredited by ICANN.[95] It had been assumed that with the creation of ICANN, NSI would recognise ICANN as the relevant regulatory body for the gTLD registry in place of the DOC. In fact as all of NSI's contracts were with the DOC and not ICANN, NSI refused to so recognise the fledgling regulator. This was the obvious and prudent business strategy, for as is noted by Milton Mueller, were NSI to do otherwise '[it would] cede control over the asset upon which its entire business had been built'.[96] There was little ICANN could do to bring the errant registrar into line. It removed much of NSI's voting rights from ICANN's policy-making bodies, but this had little effect on NSI, which felt they were unlikely to be listened to in any event. NSI went on the offensive. They refused to recognise ICANN's accreditation process, making it impossible for ICANN to extend its list of accredited registrars and they attacked ICANN's funding sources, managing to substantially undermine ICANN's financial stability in the summer of 1999, and eventually forcing Vint Cerf to secure a loan of $500,000 from his employer, MCI, to keep the organisation afloat. With ICANN's future looking decidedly uncertain, supporters of the organisation lobbied the DOC for support. A series of round-table discussions between ICANN, NSI and the DOC ensued and in September 1999, a series of agreements were put in place to settle the parties' differences.[97] These agreements provided that NSI would recognise ICANN's authority in relation to the root and the DNS, enter into a registry accreditation agreement with ICANN, and would provide ICANN with $1.5 million in badly needed financial support. In return ICANN would license NSI to manage the gTLD Registry for four years, with a guaranteed renewal for another four years should NSI divest its registrar functions within 18 months.[98] Thus NSI would retain control over the 'A' root server and would be the only entity in a position to record new entries in the gTLD database. To retain this control NSI had to agree a substantially reduced wholesale registry price of only $6 per name, per year, down from the previously agreed

94 ICANN, *ICANN Names Competitive Domain-Name Registrars*, press release, 21 April 1999, available at: *www.icann.org/announcements/icann-pr21apr99.htm.*
95 Muller, see above, fn 93, p 194.   96 Ibid.
97 *Approved Agreements among ICANN, the US Department of Commerce, and Network Solutions Inc*, 10 November 1999, available at: *www.icann.org/nsi/nsi-agreements.htm.*
98 It should be noted that when the agreements were renegotiated in 2001, NSI, now owned by Verisign, agreed to give up control of the. org Registry in return for a new commercial agreement that allowed it to retain its registry business while retaining control over the .com Registry.

$9 per name, per year and they had to agree to only accept registrations from ICANN-accredited registrars. Thus ICANN gained regulatory control of the gTLD Registry, but at the cost of proprietary control. ICANN now had the authority to regulate NSI through their accreditation agreement, and as such had direct authority over its role as a registrar. In doing so though they had ceded direct control over the registry function, the day-to-day management of the root DNS, to NSI. In addition, as they could not hope to take back this control at any point in the short to medium term, they could not plan any changes to the registry process without the agreement of NSI. Thus although NSI's monopoly over the registration of gTLDs was removed, it was done so at a heavy cost to both the DOC and ICANN. ICANN in particular had had to concede substantial ground to NSI, including crucially the registry function, to gain regulatory control over the gTLD system, and their battles were not finished yet.

NSI were not the only body following ICANN's emergence. As already stated, trade mark holders had become increasingly concerned about the actions of cybersquatters, typosquatters (individuals who would register common misspellings of well-known names like coco-cola.com), and domain name hijackers (individuals who would register well-known names and then use them to peddle porn or similar as in the famous whitehouse.com case).[99] With the emergence of the gTLD-MoU, intellectual property holders had lobbied for strong trade mark protection in any new management regime for the DNS root. They argued that a centralised system of control allowed for a centralised system of policing cybersquatters and the like. Some proposed an extension of trade mark rights into the DNS, suggesting that the DNS recognise their right in that name or mark by excluding others from using the name or mark in the DNS, however it may be used.[100] Although this was never successful, it was, as noted by Milton Mueller, an attempt by the trade mark holders to shift the transaction costs of trade mark protection from the trade mark holders to the registry and registrar markets;[101] unfortunately this has not been the only attempt on the part of trade mark holders to shift these overheads onto others as we shall see. The DOC placed the trade mark holders' concerns at the centre of its proposed reform, and made trade mark dispute resolution a key part of ICANN's mandate.[102] As part of this, the

---

99 Pelline, J, 'Whitehouse.com Goes to Porn', *CNET News*, 5 September 1997, available at: *http://news.com.com/2100–1023–202985.html?legacy=cnet*.
100 Mueller, see above, fn 93, p 190.   101 Ibid.
102 Para 9(d) of the *Memorandum of Understanding between the U.S. Department of Commerce and Internet Corporation for Assigned Names and Numbers of 25 November 1998*, states that ICANN shall take into account the following: 'Recommendations regarding trademark/domain name policies set forth in the Statement of Policy; recommendations made by the World Intellectual Property Organization (WIPO) concerning: (i) the development of a uniform approach to resolving trademark/domain name disputes involving Cyberpiracy;

DOC authorised the WIPO to create a set of policy recommendations for ICANN. WIPO is not an unbiased body.[103] Although formally WIPO is a specialised agency of the United Nations (UN), its role is entirely to develop and strengthen the legal and administrative procedures to protect and enforce intellectual property rights (IPR), as can clearly be seen in its 'vision' section of its Medium-Term Plan for WIPO Program Activities[104] where it states that: 'The main objectives of the Medium-Term Plan, as expressed in the past remain constant: maintenance and further development of the respect for intellectual property throughout the world. This means that any erosion of the existing protection should be prevented, and that both the acquisition of the protection and, once acquired, its enforcement, should be simpler, cheaper and more secure.' Not surprisingly, when offered the opportunity to revise the regulatory regime of the root DNS and registry, WIPO took full advantage. In June 1998, in immediate response to the DOC White Paper published on 5 June, WIPO undertook an international consultation to develop recommendations concerning the intellectual property issues associated with internet domain names, including domain name dispute resolution. When this initial consultation closed WIPO collated the opinions expressed and produced an interim report and proposals.[105] This report, rather than reading like a balanced proposal for a policy on the regulation of the DNS while providing protection for intellectual property interests, read like a 'wish list' for intellectual property holders. The report proposed:

(a) A comprehensive scheme for alternate dispute resolution for all intellectual property disputes that might involve a domain name. These rights to include non-trade-mark-related rights such as 'rights of personality'[106] The alternate dispute resolution system proposed being a mandatory online international arbitration system;
(b) That globally 'famous' trade marks be made unavailable for registration in all gTLDs, except in the hands of the rights holder;

(ii) a process for protecting famous trademarks in the generic top level domains; (iii) the effects of adding new gTLDs and related dispute resolution procedures on trademark and intellectual property holders; and recommendations made by other independent organizations concerning trademark/domain name issues.'

103 Milton Mueller notes that it is 'entirely beholden to intellectual property owners'. Mueller, see above, fn 93, p 190.
104 Director General of WIPO, *Medium-Term Plan for WIPO Program Activities – Vision and Strategic Direction of WIPO*, A/39/5, 21 July 2003, available from *www.wipo.int/about-wipo/en/dgo/pub487.htm.*
105 WIPO, *The Management of Internet Names and Addresses: Intellectual Property Issues*, RFC–3, 23 December 1998, available at: *http://arbiter.wipo.int/processes/process1/rfc/3/interim2.html.*
106 The right of personality is a controversial principle, which gives persons, including politicians, actors, and other famous people, special rights over the use of their names.

(c) That no distinction to be made between commercial and non-commercial uses of domain names;

(d) That failure to fully and accurately complete contact information when registering a domain name would be grounds for forfeiting that name, even without evidence of abuse; and

(e) Databases of registrant's contact details, including the WHOIS database, be made cheaply and easily accessible to allow intellectual property rights holders to identify those in violation of their rights.

Unsurprisingly the response from the internet community was powerful and critical. Professor Michael Froomkin of the University of Miami, noted that the WIPO proposal was: 'biased in favour of trademark holders; fails to protect fundamental free-speech interests including parody, and criticism of corporations; provides zero privacy protection for the name, address and phone number of individual registrants; and creates an expensive loser-pays arbitration process with uncertain rules that will intimidate persons who have registered into surrendering valid registrations'.[107] Ellen Rony, co-author of the *Domain Name Handbook*, noted that 'the proposed WIPO recommendations for settling domain name conflicts are far too broad and overreaching. WIPO creates a whole new system of administrative law to regulate domain names on behalf of a special group of Internet users – trademark owners,'[108] while Adam Woeger, the owner of the salvation.com domain name asked, 'Is WIPO trying to invent new international law regarding trademarks and domain name disputes, in a world where only national trademarks are issued? What right does WIPO have to do this?'[109] The reactions from the internet community forced WIPO to reconsider its policy. When its final report was published on 30 April 1999, WIPO had substantially retreated its position. Now the alternate dispute resolution (ADR) system would only be applicable to those cases where the intellectual property rights holder could establish that the domain name registration was 'abusive,'[110] a term defined in paragraph 171 of the report as occurring when 'the domain name is identical or misleadingly similar to a trade or service mark in which the complainant has rights; and (i) the holder of the domain name has no rights or legitimate interests in respect of the domain name; and (ii) the domain name has been registered and is used in bad faith', and protection would be afforded to

107 Froomkin, M, *A Critique of WIPO's RFC 3*, 1999, available at: *www.law.miami.edu/~amf/ critique.htm*.

108 Rony, E, *Comments on the World Intellectual Property Organization Interim Report: RFC–3*, 1999, available at: *www.domainhandbook.com/rfc3.html*.

109 Email from Robert Adam Woeger to WIPO. Available at: *http://arbiter.wipo.int/ processes/process1/rfc/dns_comments/rfc3/0018.html*.

110 WIPO, *The Management of Internet Names and Addresses: Intellectual Property Issue, Final Report of the WIPO Internet Domain Name Process*, 30 April 1999, paras 152–228. Report available from: *http://arbiter.wipo.int/processes/process1/report/pdf/report.pdf*.

domain name holders against spurious or bad faith challenges to their name by protecting against the practice of 'reverse domain name hijacking'.[111] For ICANN, protection of trade mark interests was second only to the creation of competition in gTLDs in terms of priority. With ICANN in control of the root server, problems with NSI notwithstanding, and the register accreditation system in place, it was in position to implement the WIPO proposals. On 27 May 1999, the ICANN Board adopted a resolution to receive the recommendations of the WIPO final report. This led to the creation of an ICANN working group to study the WIPO recommendations. By 29 July, the working group had compiled its final report. The report recommended establishing a uniform domain-name dispute-resolution policy (UDRP) for all registrars, a proposal that was adopted by the Domain Names Supporting Organisation (DNSO) Names Council on 4 August. The report was then sent to the main ICANN Board for implementation. On 26 August, the ICANN Board accepted the DNSO recommendation and by 24 October the ICANN UDRP was in place. Thus from the publication of the WIPO report on 30 April 1999, it took only 176 days for ICANN to implement its recommendations. The ADR component of the Policy was not to be supplied by ICANN itself. Rather they would license 'approved dispute-resolution service providers' who would supply the panellists to hear claims and who would manage the administration of complaints. With the new procedure 'going live' on 1 December, ICANN announced on 29 November that the first approved dispute-resolution service provider would be WIPO, who have since been joined by the National Arbitration Forum,[112] the Disputes.org/eResolution consortium,[113] the CPR Institute for Dispute Resolution,[114] and the Asian Domain Name Dispute Resolution Centre.[115] The first UDRP claim was raised on 9 December 1999; the domain name in dispute being worldwrestlingfederation.com, an action that led to success for the claimant on 14 January 2000.[116]

Since that first ICANN UDRP decision was issued in January 2000, the UDRP has handled over 10,000 complaints involving over 18,500 domain names.[117] This makes the UDRP by far the most successful and far-reaching

111 Reverse domain name hijacking occurs when a rights holder attempts to use their rights in a name or mark to obtain, without due cause, the domain name of another. It is discussed in the report, paras 323–7, 'Resort to Defensive Practices'.
112 Approved on 23 December 1999.
113 Approved on 1 January 2000. eResolution stopped accepting proceedings under the UDRP on 30 November 2001.
114 Approved on 22 May 2000.   115 Approved on 28 February 2002.
116 *World Wrestling Federation Entertainment Inc v Michael Bosman* Case No. D99–0001, available at: *http://arbiter.wipo.int/domains/decisions/html/1999/d1999–0001.html.*
117 A survey of the decisions reported on the ICANN List of Proceedings under the Uniform Domain Name Dispute Resolution Policy website (*www.icann.org/udrp/proceedings-list-number.htm*) on 10 May 2005 revealed that 10,733 proceedings had been raised involving 18,575 domain names.

aspect of ICANN's functions. Despite its popularity though, the UDRP has been heavily criticised. The most vociferous critic of the UDRP has probably been Professor Milton Mueller of Syracuse University. In his book, *Ruling the Root*, he describes the UDRP as: 'heavily biased in favour of complainants. It allows the trademark holder to select the dispute provider, thereby encouraging dispute resolution providers to compete for the allegiance of trademark holders. The resultant forum shopping ensures that no defendant friendly service provider can survive.'[118] This may explain the pre-eminence of the WIPO UDRP service, which by 20 May 2005 had provided 6,548 UDRP decisions (61 per cent of all decisions), while collectively the other four UDRP providers had provided only 4,185 decisions (39 per cent). Another critic, Professor Michael Froomkin, focuses on the procedure's failure to comply with some of the basic principles of natural justice. He notes that the UDRP 'would have little chance of surviving ordinary "arbitrary and capricious" review, because it denies respondents minimal levels of fair procedure that participants would be entitled to expect . . . three aspects of the UDRP are particularly troubling: (1) the incentive for providers to compete to be "complainant friendly"; (2) its failure to require actual notice combined with the short time period permitted for responses; and (3) the asymmetric consequences of a decision'.[119] Elsewhere, the current author has focused upon the lack of training or experience of UDRP panellists, noting that 'almost half of all panellists employed by the two major UDRP Providers[120] are untrained and inexperienced in adjudication'[121] and the potential bias of panellists noting: 'the WIPO UDRP panel contains a high proportion of Intellectual Property practitioners. Of the 193 WIPO panellists currently practicing within the legal profession, 110 (57 per cent) list a specialism in intellectual property. In addition, of the 41 academic lawyers listed, 22 (53.7 per cent) are listed as intellectual property professors or lecturers. Although it is to be expected that a high proportion of UDRP panellists would be experienced in intellectual property law given the nature of the disputes in question, and although there is no claim here made of individual bias by panellists in favour of intellectual property rights holders, for those panellists involved in the practice of IP law it may be difficult to maintain neutrality as the major aspect of their full-time vocation is the protection of IP rights from erosion and this might be expected to mean that certain "habits of thought" are prevalent'.[122] The UDRP may therefore be classified as a policy that has been successful and popular with the community at large, but one which is

---

118   Mueller, see above, fn 93, p 193.
119   Froomkin, M, 'Wrong Turn in Cyberspace: Using ICANN to Route Around the APA and the Constitution', 2000, 50 *Duke Law Journal* 17, p 136.
120   The two major UDRP providers being the WIPO and the National Arbitration Forum.
121   Murray, A, 'Regulation and Rights in Networked Space', 2003, 30 JLS 187, p 203.
122   Ibid, p 216.

controversial. But with much of ICANN's public image tied to the UDRP, ICANN had to do everything possible to support the UDRP and its service providers. But more trouble was to come.

ICANN's legitimacy

In the summer of 2000 events were to cast a further shadow over ICANN's regulatory legitimacy and its ability to control and manage the root and the DNS. ICANN's original Bylaws of 6 November 1998 required that the original board members terminate their service on 30 September 1999, and that in their place new 'At Large' directors should be elected. To allow for these elections to occur, ICANN created a Membership Advisory Committee to design and implement a membership system. This small, but highly efficient group completed their task by May 1999, proposing to the board a highly democratic model for elections.[123] They proposed that any internet user could register as a member of ICANN's At Large community, that they should be able to do so without payment of a fee and that their membership would be renewable annually. They further proposed that five board directors be elected annually on a regional basis by the At Large membership, each member having one vote in these elections. The elections were to take effect once 5,000 At Large members had been registered. The board seemed to be less keen. At a meeting of the board on 27 May 1999, they resolved that elections were likely to be administratively complex and expensive and that the costs should be borne by the membership. With no clear procedure for electing new At Large directors in place, the original members of the board re-appointed themselves for a further 12 months at their Santiago meeting on 26 August 1999.[124] Under the terms of the ICANN Bylaws though, they could only extend their tenure to the 30 September 2000, so the issue of selecting At Large directors remained urgent. The board chose to abandon all of the Membership Advisory Committee's recommendations. They suggested replacing the direct election of At Large directors with the creation of an At Large council, made up of members selected by the At Large community, who would in turn elect At Large board members. The failure of these proposals to protect the principle of direct election of At Large directors attracted outrage among the internet community, and facing the prospect of a grass roots rebellion at the ICANN Cairo meeting in March 2000, the board acquiesced and promised that five At Large directors would be elected on a regional basis in Autumn 2000, each to serve a two-year term.

123 ICANN, *Principles of the At-large Membership: Recommendations of the Membership Advisory Committee*, 5 May 1999, available at: *www.icann.org/berlin/membership_rec.htm*.
124 ICANN, Resolutions Approved by the Board at the Santiago Meeting, 26 August 1999, available at: *www.icann.org/santiago/santiago-resolutions.htm*.

The level of dissatisfaction within the internet community with the board, and ICANN's policies soon became clear. When the results of the At Large elections were finally announced on 11 October 2000, they reflected a devastating rejection of the current board and its policies. In the North American region, seven candidates stood. Four were nominated by ICANN (Donald Langenberg,[125] Lyman Chapin,[126] Harris Miller[127] and Lawrence Lessig) and three were nominated by the members (Emerson Tiller,[128] Barbara Simons[129] and Karl Auerbach[130]). Of the seven nominees only Donald Langenberg, Lyman Chapin and Harris Miller could be described as supportive of the ICANN board and its current policies; the others were all critical, Lawrence Lessig included, with Karl Auerbach being by far the most critical.[131] When the votes were counted Karl Auerbach had won a convincing majority over Barbara Simons and Lawrence Lessig with Donald Langenberg, Lyman Chapin and Harris Miller collectively only mustering 389 votes, compared to Auerbach's winning count of 1,074 votes. A similar picture emerged in the European constituency. There, seven candidates stood – five nominated by ICANN (Oliver Popov,[132] Maria Livanos Cattaui,[133] Alf Hansen,[134] Olivier Muron[135] and Winfried Schüller[136]) and two were nominated by the members (Jeanette Hofmann[137] and Andy Mueller-Maguhn[138]). All five of the ICANN nominees were supportive of the ICANN board and its current policies, with

125 Professor Donald Langengberg was Professor of Physics and Chancellor of the University of Maryland.
126 Lyman Chapin was Chief Scientist at BBN Technologies, the company that built the original IMPs.
127 Harris Miller was President of the Information Technology Association of America (ITAA).
128 Emerson Tiller was Associate Professor of Business, Technology and Law and Co-director of the Center for Business, Technology and Law at the University of Texas at Austin.
129 Barbara Simons was President of the Association for Computing Machinery.
130 Karl Auerbach was a senior researcher in the Advanced Internet Architecture group in the Office of the Chief Strategy Officer at Cisco Systems.
131 Auerbach stood on a platform of reform, including reform of ICANN's procedures and personnel. See the Auerbach campaign website at: *www.cavebear.com/ialc/platform.htm*.
132 Oliver Popov was a professor of Computer Science at the Institute of Informatics, University of St Cyril and Methodius, Skopje.
133 Maria Livanos Cattaui was the Secretary-General of the International Chamber of Commerce.
134 Alf Hansen was Director of UNINETT FAS A/S in Trondheim, Norway. UNINETT FAS is the host for NORID, the Norwegian Registration Service for Internet Domain Names.
135 Olivier Muron was in charge of IP technology and electronic commerce at France Telecom.
136 Winfried Schüller was Director of International IP Services at Deutsche Telekom.
137 Jeanette Hofmann was Program Leader of the Research Unit on Internet Governance at Wissenschaftszentrum Berlin für Sozialforschung (WZB) (The Social Science Research Centre, Berlin).
138 Andy Mueller-Maguhn was a professional journalist and member of the Chaos Computer Club, one of the biggest and most influential hacker organisations.

Jeanette Hofmann being critical and Andy Mueller-Maguhn, highly critical. Again, the electorate spoke with Mueller-Maguhn winning a landslide – 5,948 votes compared to his nearest competitor Hofmann with only 2,295. The five ICANN nominees only managed to muster 3,066 votes between them. When the results came out, Andy Mueller-Maguhn noted that 'the Internet community consensus that ICANN had been claiming since its inception seemed not to exist'. The ICANN board reacted in unfortunately predictable fashion to the election results. The board created a new 'executive committee', which excluded the At Large directors and passed to it much of the board's executive powers. They also amended the ICANN Bylaws to exclude At Large directors from key processes such as the selection of new gTLDs.[139] The At Large directors were excluded from key decisions to such an extent that Karl Auerbach began to keep a 'decision diary' on his website listing decisions he had been party to and those he had not.[140] Relations between Auerbach and the ICANN board eventually became so eroded that on 18 March 2002, Karl Auerbach petitioned the Superior Court for the County of Los Angeles under § 6334 of the California Corporations Code requesting a 'Writ or Order to the Respondent, ordering and directing the Respondent immediately to make available to Petitioner for inspection and copying all corporate records of the Respondent which Petitioner sets forth in this Petition, or which Petitioner may request access to from time to time.'[141] Auerbach's claim was that he had been denied access to key financial data including the Corporation's 'General Ledger' of financial data from November 1998 to present. Stuart Lynn, ICANN's CEO, initially refused to allow him access, then agreed, but on condition that Auerbach visit ICANN's offices to make the inspection and that:

(1) Although Auerbach had the right to be accompanied by his attorney or other advisor, Lynn reserved to himself the right to veto the person(s) selected by Auerbach;

(2) Although Auerbach would be allowed to inspect paper copies of the

---

139 The authorisation of new generic TLDs was a key ICANN policy decision. Almost immediately upon its formation ICANN began work on this by setting up an open working group on TLDs. This working group reported in March 2000 that ICANN should authorise between six and ten new gTLDs, followed by an evaluation period. The board accepted this proposal at its Yokohama meeting in July 2000, but also decided to remove the new At Large directors from the decision-making process, ensuring that the final seven new gTLDs were selected by the original board members. The seven selected were .biz, .info, .pro, .name, .aero, .coop and .museum, all proposals which had been put forward by groups that Milton Mueller describes as 'established, politically connected insiders'. Mueller, see above, fn 93, p 203.

140 This may be found at *www.cavebear.com/icann-board/diary/index.htm*.

141 Taken from Karl Auerbach's petition. Available at: *www.icann.org/legal/auerbach-v-icann/petition–18mar02.pdf*.

records requested, he would not be given electronic copies as requested of at least some of them; and

(3) Only after Auerbach had inspected the records could he designate those for which he wanted copies made. Lynn would then consider Auerbach's 'request' for copies with the advice of and in consultation with the Audit Committee, at which point copies of the records might or might not be provided to Auerbach.[142]

Karl Auerbach, believing these conditions to be unacceptable chose to petition the court. In their reply, ICANN claimed the action stemmed from Auerbach's refusal to inspect the documents without conditions as to access. They claimed that: 'a director's right to inspect documents is not "absolute" in the sense that a corporation cannot place into effect reasonable procedures for inspections that safeguard the interests of the corporation. ICANN's procedures are, in fact, completely reasonable and prudent under the circumstances, and they would not limit Auerbach's ability to inspect ICANN's documents.'[143] The Petition was finally decided, in favour of Karl Auerbach, on 5 August 2002: the Superior Court ordering ICANN to make available 'all corporate records of the Respondent'.[144]

Auerbach's dispute on access may have been won by the At Large director, but the breakdown in trust between ICANN's board and its At Large directors, and it seems the wider internet community, was now such that when the five elected At Large directors completed their terms of office at the 26 June 2003 (Montreal) board meeting, no new elections to find their replacements were held. Three of the At Large directors, including Mueller-Maguhn and Auerbach were retired immediately, while two – Masanobu Katoh, the Asia-Pacific representative, and Ivan Moura Campos the representative of the Latin America and Caribbean region – were kept on, and were joined by new directors who had been selected by the newly formed Nominating Committee (NomCom), which is now responsible for the selection of all ICANN directors except the president and those selected by ICANN's supporting organisations. By creating NomCom, the board had dispensed with the problem of 'maverick' directors, but at a high cost. The legitimacy that ICANN claimed to have was now clearly tarnished. It had flirted with democracy and had found that the community it served seemed to be of a different view to the board. Its response had been reminiscent of an Eastern-bloc dictator: it had taken power back to the centre and had

---

142 See Auerbach, above, fn 141.
143 Taken from ICANN's Reply dated 17 April 2002, available at: *www.icann.org/legal/auerbach-v-icann/answer–17apr02.pdf*.
144 Taken from the Judgment of the Superior Court of California at Los Angeles, available at: *www.eff.org/Infrastructure/DNS_control/ICANN_IANA_IAHC/Auerbach_v_ICANN/20020807_auerbach_judgment.pdf*.

appointed a Politburo drawn from those sympathetic to the board and its policies. ICANN's perceived lack of legitimacy, and perhaps more pressingly, its lack of responsiveness to external demands from the internet community and other key stakeholders, including governments and other regulatory bodies, has now placed ICANN firmly in the sights of regulatory modernisers.

Enter the international community

As with all regulators and regulatory settlements, the actions of ISOC and ICANN were not developing within a regulatory vacuum and the perceived regulatory failures of ICANN in particular were attracting attention from the international community. The World Summit on the Information Society (WSIS) is the highest profile event to date to deal with the threats and opportunities offered by information and communications technology (ICT). The need for a UN Summit[145] on this issue was first identified by the ITU in 1998, when by Resolution 73 of the ITU Plenipotentiary Conference in Minneapolis, they noted that telecommunications were playing an increasingly decisive and driving role at the political, economic, social and cultural levels and called upon the UN: 'to ask the Secretary-General to coordinate with other international organizations and with the various partners concerned (Member States, Sector Members, etc.), with a view to holding a world summit on the information society.'[146] This request was heard at the ninetieth plenary meeting of the General Assembly of the United Nations in December 2001, where the General Assembly accepted and endorsed a proposal from the ITU that a WSIS be convened, and instructed the Secretary-General of the UN to 'inform all heads of State and Government of the adoption of the present resolution.'[147] The WSIS was to take place in two phases – the first phase taking place in Geneva on 10–12 December 2003 and the second phase taking place in Tunis, on 16–18 November 2005. The objective of the Geneva phase was to develop and foster a clear statement of political will and take concrete steps to establish the foundations for an information society for all, reflecting all the different interests at stake. The objective of the second phase was to put the Geneva 'Plan of Action' into effect and to find solutions and reach agreements in the fields of internet governance, financing mechanisms, and follow-up and implementation of the Geneva and Tunis documents.

---

145   United Nations Summits are designed to put long-term, complex problems like poverty and environmental degradation at the top of the global agenda. They are designed to provide leadership and to mould international opinion and to persuade world leaders to provide political support.

146   Resolution 73, available at: *www.itu.int/wsis/docs/background/resolutions/73.html.*

147   Resolution adopted by the General Assembly [on the report of the Second Committee (A/56/558/Add.3)] 56/183. World Summit on the Information Society, 21 December 2001.

While it is too early to gauge the success, or otherwise, of the WSIS,[148] there is little doubt that it has begun a new chapter in the discourse in global communications and media governance. The WSIS invited heads of state/government, international non-government organisations (NGOs) and civil society representatives[149] to contribute to a series of preparatory meetings (PrepComms) and to the Geneva and Tunis rounds on a series of issues ranging from the digital divide,[150] to freedom of expression, network security, unsolicited commercial communications (SPAM) and protection of children.[151] Central to the WSIS programme was though the issue of internet governance.

WSIS envisaged a 'people-centred, inclusive and development-orientated Information Society where everyone can create, access, utilize and share information and knowledge, enabling individuals, communities and peoples to achieve their full potential in promoting their sustainable development and improving their quality of life'.[152] These principles were at odds with the commonly held view of internet governance as a western-led process dominated by the Government of the United States and (mostly US-based) NGOs such as ICANN, with developing nations largely absent from the process. As a result WSIS, it appeared, would have to tackle, head-on, the dominance of western industrialised nations and in particular, ICANN, in managing the root server system and the addressing protocols of the logical infrastructure layer. Discussion as to how this was to be achieved began in the PrepComms. In these meetings numerous views were expressed about what was and was not 'internet governance', and the public policy involved. Some developing nations noted that they were unable to participate in many of the

---

148 Many early commentators on WSIS have been critical of its lack of effect or ambition. See eg Hamelink, C, 'Did WSIS Achieve Anything At All?', 2004, 66 *Gazette: The International Journal for Communication Studies* 281 (Referring to the Geneva Round), c/f Raboy, M, 'The World Summit on the Information Society and Its Legacy for Global Governance', 2004, 66 *Gazette: The International Journal for Communication Studies* 225; Diab, K, 'Walk First then Surf', 772 *Al-Ahram Weekly*, 8–14 December 2005 (Referring to the Tunis Round).

149 In UN parlance, civil society encompasses all those who are not part of government, private enterprise or intergovernmental organisations – in other words private individuals.

150 The 'digital divide' reflects the technology gap that has opened up between technology rich western States and technology poor African and Asian States, and on the growing divide within States between the professional classes with stable and fast internet access and the working class, in particular immigrant communities, where access may be unstable, slow and difficult to obtain. See Norris, P, *Digital Divide: Civic Engagement, Information Poverty and the Internet Worldwide*, 2001, Cambridge: CUP; Warschauer, M, *Technology and Social Inclusion: Rethinking the Digital Divide*, 2004, Cambridge, MA: MIT Press.

151 For a discussion of the WSIS see Raboy, M and Landry, N, *Civil Society, Communication And Global Governance: Issues from the World Summit on the Information Society*, 2004, Bern: Peter Lang.

152 WSIS, Declaration of Principles, Geneva, 12 December 2003, Principle 1.

decision-making processes regarding management of the root and allocation of IP numbers and associated domain names. Particularly, they felt unable to manage resources they believed they had a right to manage, predominantly a sovereign right in the case of country code top-level domains (ccTLDs). Others, predominantly the US, called for the principle of private sector involvement and investment to be enshrined. In a final PrepComm briefing on 3 December 2003, US Ambassador David Gross outlined what he called the 'three pillars' of the US position. These were: (1) as nations attempt to build a sustainable ICT sector, commitment to the private sector and rule of law must be emphasised so that countries can attract the necessary private investment to create the infrastructure; (2) the need for content creation and intellectual property rights protection in order to inspire ongoing content development; and (3) insuring security on the internet, in electronic communications and in electronic commerce.[153] Thus when the Geneva Summit got under way the PrepComms had failed to produce agreement on the future development of internet governance. Although committed to a principle of multi-stakeholder agreement, many developing nations, including China, Brazil and most Arab States saw the US commitment to private sector initiatives as a barrier to progress while the US and others, including the EU, Japan and Canada, feared that some governments wished to have a greater say in internet governance purely as a vehicle for censorship or content management. As a result agreement in Geneva proved impossible. Instead, it was noted that: 'governance issues related to the internet are a complex challenge which needs a complex answer and which has to include all stakeholders – civil society, private industry and governments. No single body and no single stakeholder group alone is able to manage these challenges. This multi-stakeholder approach should be the guiding principle both for the technical co-ordination of the internet, as well as for broader public policy issues related to Cyberspace in general.'[154] To give effect to this recommendation, WSIS put together a Working Group on Internet Governance (WGIG) to report to the Tunis conference with recommendations. The group, chaired by Nitin Desai, Special Adviser to the Secretary-General for the WSIS, met four times between Geneva and Tunis, and published their final report on 18 July 2005.[155]

The group was asked to carry out their work under three broad heads: (1) to develop a working definition of internet governance; (2) to identify the public policy issues that are relevant to internet governance; and (3) to

153   See US Outlines Priorities for World Summit on the Information Society, available at: *http://usinfo.state.gov/xarchives/display.html?p=washfile-english&y=2003&m=December&x=20031203163730retropc0.0570032&t=usinfo/wf-latest.html*.
154   World Summit on the Information Society, *Visions in Process: Geneva 2003 – Tunis 2005*, p 41, available at: *www.worldsummit2003.de/download_de/Vision_in_process.pdf*.
155   Full details of WGIG may be found at its website: *www.wgig.org/*.

develop a common understanding of the respective roles and responsibilities of governments, existing international organisations and other forums, as well as the private sector and civil society in both developing and developed countries.[156] In dealing with the first the group suggested the following working definition:

> Internet governance is the development and application by Governments, the private sector and civil society, in their respective roles, of shared principles, norms, rules, decision-making procedures, and programmes that shape the evolution and use of the internet.[157]

This shows a qualified victory for supporters of the broad definition of governance. Some delegates had suggested that a narrow, technical, view of internet governance as only those technical issues carried out by ICANN and ISOC. This view is clearly rejected in paragraph 12 of the report where it is noted that: 'Internet governance includes more than internet names and addresses, issues dealt with by the Internet Corporation for Assigned Names and Numbers (ICANN): it also includes other significant public policy issues, such as critical internet resources, the security and safety of the Internet, and developmental aspects and issues pertaining to the use of the internet.' This though, is not a complete victory for supporters of the broad definition, as many wished to have it extended to cover issues such as appropriateness of content, taxation and e-commerce, issues that appear to be precluded by the definition agreed. From this base the group then moved on to its second head of study and listed 13 public policy issues 'of the highest priority'.[158] With this achieved the group than made its critical recommendations on the respective roles and responsibilities of governments, existing international organisations and other forums. It recommended that governments were to drive public policy-making and co-ordination and implementation, as appropriate, at the national level, and policy development and co-ordination at the regional and international levels.[159] This was to include development of best practices, capacity-building and promoting research and development. The private sector meanwhile was called upon to develop policy proposals,

---

156 Taken from WGIG, *Report of the Working Group on Internet Governance*, Château de Bossey, 18 June 2005, para 5.

157 Ibid, para 10.

158 These were: (1) Administration of the root zone files and system; (2) Interconnection costs; (3)Internet stability, security and cybercrime; (4) Spam; (5) Meaningful participation in global policy development; (6) Capacity-building; (7) Allocation of domain names; (8) IP addressing; (9) Intellectual Property Rights; (10) Freedom of expression; (11) Data protection and privacy rights; (12) Consumer rights; and (13) Multilingualism. Full discussion of these may be found at paras 15–27 of the Report.

159 Ibid, para 30.

guidelines and tools for policymakers and other stakeholders, this including industry self-regulation and arbitration and dispute resolution.[160] To manage the relationship between the public and private sector (and other stakeholders) the WGIG recommended the creation of a new Internet Governance Forum (IGF), which would provide the opportunity for the free exchange of ideas between stakeholders and which would provide public policy guidance and oversight of ICANN.[161] The report suggested four models as to how the relationship between ICANN and the new IGF could be structured. The first would have seen ICANN become accountable to the IGF with the IGF in effect taking the place of the US DOC. Thus ICANN would become internationalised and would become an international NGO under oversight of a UN body. The second would have seen the IGF supplement ICANN's existing Government Advisory Committee (GAC), thus leaving ICANN substantially intact while providing a new fora for the civil society representatives to be heard in addition to the current ICANN public meetings. The third model would see the IGF replace the existing GAC as an intergovernmental forum. This would have little direct impact on ICANN, but would weaken the position of developed nations that have a disproportional representation on the ICANN board and the GAC. The final recommendation was the most radical, replacing ICANN with a 'World Internet Corporation for Assigned Names and Numbers (WICANN)', overseen by a new global IGF.[162] Understandably the four recommendations excited a great deal of debate and high-level discourse. China, Cuba and South Africa argued that the US and other wealthier nations must share power. Luisa Diogo, the Prime Minister of Mozambique stated that: 'It is a matter of justice and legitimacy that all people must have a say in the way the Internet is governed',[163] while Robert Mugabe, President of Zimbabwe warned that 'the US and allies such as the UK unreasonably insist on being world policemen on the management of the Internet, that must change'.[164] The US, though, remained resolute that no institutional change would be sanctioned and their stance seems to have paid off when just hours before the Tunis Summit opened, a deal was reached which saw the US agree to the creation of the IGF, but in a much reduced form from that envisaged by the WGIG. The mandate for the new forum is set out in Paragraph 72 of the Tunis Agenda for the Information Society[165] which states:

160   See above, fn 158, para 31.   161   Ibid, paras 35–51.
162   Ibid, paras 52–72.
163   See McCullough, D, 'US endorses Internet Governance Forum', *CNet News*, 16 November 2005, available at: *http://news.zdnet.co.uk/internet/0,39020369,39237279,00.htm.*
164   Ibid.
165   WSIS, *Tunis Agenda for the Information Society*, 18 November 2005, available at: *www. itu.int/wsis/docs2/tunis/off/6rev1.html.*

The mandate of the Forum is to:

a.   Discuss public policy issues related to key elements of Internet governance in order to foster the sustainability, robustness, security, stability and development of the Internet.
b.   Facilitate discourse between bodies dealing with different cross-cutting international public policies regarding the Internet and discuss issues that do not fall within the scope of any existing body.
c.   Interface with appropriate intergovernmental organisations and other institutions on matters under their purview.
d.   Facilitate the exchange of information and best practices, and in this regard make full use of the expertise of the academic, scientific and technical communities.
e.   Advise all stakeholders in proposing ways and means to accelerate the availability and affordability of the Internet in the developing world.
f.   Strengthen and enhance the engagement of stakeholders in existing and/or future Internet governance mechanisms, particularly those from developing countries.
g.   Identify emerging issues, bring them to the attention of the relevant bodies and the general public, and, where appropriate, make recommendations.
h.   Contribute to capacity-building for Internet governance in developing countries, drawing fully on local sources of knowledge and expertise.
i.   Promote and assess, on an ongoing basis, the embodiment of WSIS principles in Internet governance processes.
j.   Discuss, *inter alia*, issues relating to critical Internet resources.
k.   Help to find solutions to the issues arising from the use and misuse of the Internet, of particular concern to everyday users.
l.   Publish its proceedings.

This, much restricted, mandate removes any suggestion that the IGF would take direct oversight of ICANN's activities. Thus it appears that the United Nations, having played a game of diplomatic brinksmanship with the US Government during the WSIS process, failed to force institutional change. Governments aggrieved at this failure, such as the Chinese and South African delegations must now make their case all over again at the IGF meetings, the first of which meets formally for the first time in Athens from 30 October to 2 November 2006. Thus the largest summit to date on ICTs and the internet could not settle to the satisfaction of most the central issue of governance and regulation within the informational infrastructure layer. One should not assume though that this is the end of the discourse. While the US Government may have successfully shielded ICANN from this assault it seems clear that despite the WSIS process, ICANN will remain under critical assault from stakeholders of all shapes and sizes while it continues to perpetuate a system of regulatory management without any outward sign of legitimacy. Despite the failure of the WSIS to effect institutional change at this time, ICANN it seems cannot, whatever else, continue indefinitely in its present form. The success of the US negotiating team in Tunis may eventually be seen to be little more than a temporary sticking plaster. This debate is certain to be rejoined when the IGF meets later this year.

Using code to regulate in cyberspace

The analysis contained in this chapter confirms that which we would expect to find. As discussed in Chapter 2, design or architectural controls, are merely second-order regulatory instruments. They are a tool used by the first-order regulators: competition, society and hierarchy. Although a powerful regulatory tool, the effectiveness of code-based control mechanisms depends entirely upon their recognition and acceptance within these first-order regulatory environments. Thus the cyberpaternalist claim that code, left unchecked, will be utilised by business interests to close down areas of freedom, or commons, within cyberspace is found to be subject to the demands of the market. It is ultimately competition within the markets for network access and content that will decide which model of informational regulation – the model of digital property and management, or an informational commons – comes to be the standard for digital media content in the next few years. This competition is being played out on a daily basis between organisations such as the Creative Commons, the Free Software Foundation and peer-to-peer networks such as Gnutella and Kazaa and traditional content providers such as the RIAA, MPAA, the British Phonographic Institute, the Copyright Licensing Agency and suppliers such as Apple's iTunes service. The market will decide which systems become the dominant market paradigms, not code. The consumer has the power to decide this and ultimately whichever side of the competition loses the consumer vote will have to accept defeat graciously or we will be forced to witness a continual, and unedifying battle between designers of digital rights management systems and hackers, crackers and peer-to-peer systems. Equally, as we have seen, attempts to leverage or 'force' regulatory changes in cyberspace through control over the key network protocols TCP/IP seem likely to fail. The ICANN experiment seems, despite the efforts of the US negotiating team in Tunis, destined to be short-lived. Although there is no doubt that ICANN has managed to effectively leverage control over the root and the DNS in the short term through procedures such as the UDRP, it appears ICANN's lack of legitimacy within the wider community will eventually lead to its replacement. What we see occurring here is a complex interplay between competition, or market, forces in the shape of the trade mark holders, hierarchy, in the guise of the US DOC and the community. Although the competition/hierarchy modalities successfully harnessed the power of the IP code in the short term, it appears that ultimately, ICANN's lack of community recognition, or to phrase it slightly differently, its lack of a wider legitimacy, may prove to be its downfall. If this proves to be so this is an important message to all those who continue to support the cyberpaternalist school: in cyberspace the power to decide is, it seems, vested ultimately in the community. We have the power to control our destiny. Whether we seize this opportunity is up to us.

Chapter 5

Online communities

Without a sense of caring, there can be no sense of community.
Anthony J. D'Angelo

If you live in a city you will be aware of the need to comply with social conventions. Such conventions are a necessary form of democratic control, developed for the good of the community, and a necessary part of modern city life. As cities become larger, develop greater urban density and become more congested, the inhabitants of these 'monstrous ant hills' seem to be forever destined to give up more of their personal space and privacy to service the good of the community.[1] A good example of these conventions in action may be found in the etiquette of travel and transportation. Transport systems are the arteries of a modern city: to live, cities must transport large numbers of people from one area of the city to another every day. To realise this, modern cities use complex co-ordinated transport strategies, which in most cities are a combination of rail, road and light rail systems, such as trams and underground railways. The success or failure of a cities' transport strategy may seem to rest with the city planners and with the transport providers, but in fact a considerably more complex relationship is at the heart of modern intra-city transport systems – a key component which is the application of widely accepted social conventions.

I live in Greater London. London is a sprawling city that is served by a complex urban transport network, which makes use of road, rail, light railway, river buses, trams and an underground network. It has one of the most diverse and ageing transport networks in the world, which today is routinely in operation in excess of its designed capacity. The only way Transport for London (the planning and management organisation for London transportation) can

---

1 The phrase 'monstrous ant hill' is used by William Wordsworth to describe London in 'Residence in London', *Prelude* 7, 149–50, 1850, from *The Complete Poetical Works of William Wordsworth*, 1888, London: Macmillan.

keep the city moving is with the assistance of its customers. A complex set of rules, known as 'Tube etiquette', has been developed by the community and with the support of Transport for London. It is these community-based rules that allow the London Underground network to function smoothly (at least most of the time) and provide a degree of comfort during journeys taken on the Underground, despite its antiquated network and ageing physical infra-structure. These 'rules' include: standing to the right on escalators so that people in a hurry can walk on the left; standing clear of the closing doors to allow the train to proceed without delay; letting passengers off the train before you board; giving up your seat to anyone who might need it more than you, in particular the elderly and the pregnant; moving down into the carriage to use all available space; not playing your personal stereo at high volume and not invading your neighbour's personal space by spreading legs or personal belongings.[2] Although strongly supported, and widely advertised, by Transport for London, these rules are not of a hierarchical nature: they have no external penalties and have not been developed by the regulators or suppliers of the underground system. They have instead developed organ-ically from among the body of underground users and have a system of social sanctions or stigma that are used to ensure compliance. Social conven-tions similar to these control all aspects of our everyday lives, whether we live in cities or not. In any area of high-urban density there will be a set of conventions designed to ensure good neighbourly conduct.

My grandparents used to live in a tenement block in Edinburgh.[3] There were eight flats in the tenement and there were strict rules to be followed. There was a shared drying area to the rear of the block and each of the flats had specific times during the week when they could make use of this. If your washing was hanging out during the time allocated to your neighbour you could expect to have it taken down and to receive a strong admonishment. Similarly, each flat took turns, in strict rotation, to clean all the common areas of the block and anyone who missed their turn could expect a similarly sharp rebuke. Nowadays, many of these old rules have expired and in their place a new set of social conventions have developed. Simple rules such as not playing your music or television too loud, especially in the late evening and overnight; not leaving rubbish in common areas and not allowing alarms (car or property) to sound continuously have replaced the older rules. Like the transportation rules discussed previously these rules are in many cases

---

2 To read more on this visit Transport for London's Tube Etiquette page at: *www.tfl.gov.uk/ tube/using/useful-info/etiquette.asp*.

3 A tenement is an apartment building built specifically to house multiple working-class families. Tenements were built in Scottish cities throughout the Victorian period to house the increasing numbers of families moving to the cities from the countryside, following the Industrial Revolution. Many tenements still survive in Scottish cities due to their efficient design.

supported, and in some cases even enforced, by hierarchical controls that have crystallised the social conventions. I live in a modern managed development: the terms of the lease for this development state that music should be turned down after 9.00 pm, that residents should keep the development tidy and in good repair and that no-one is to install a satellite dish. Similarly, my local authority has a policy to deal with noise pollution and will investigate any complaints of noise pollution emanating from noisy neighbours. These are hierarchical controls that have grown out of previous social conventions. These hierarchical policies can though only be invoked when an abuse occurs. Simple good neighbourliness, that is proactive rather than reactive regulation, is entirely in the hands of the community and it is the community that must act to enforce its standards against errant members. These are two simple, everyday examples of how communities self-regulate. If you think of examples from your daily routine you will discover dozens of similar examples of everyday self-regulation within communities. Your appearance offers one such simple example. Do you ever stop to consider what you are wearing? Do you question why people replace clothes that are still in good repair, but which have gone out of fashion? Or why do people pay out hundreds of pounds on designer labels when cheap alternatives may be easily sourced in the high street? The answer to all these is of course, community, or peer, pressure. People are willing to spend hundreds of pounds on a Fendi bag, Matthew Williamson dress or a pair of Manolo Blahnik shoes because they feel under intense pressure to live up to the expectations of their peers. They fear that should they not meet the high community fashion standards expected today their peers will mark them out as some sort of failure, meaning some people are willing to take on heavy debts to create an image of success. These standards of fashion and style are communicated via high-street stores, the media and through everyday interaction. It is not always clear where the lead comes from, although it is usually assumed that couture fashion shows are giving a lead. Wherever these standards develop though, they clearly communicate to us how the community expects us to maintain our appearance.[4]

Communities, unlike design, which we looked at in the previous chapter, control through consensus. A community may only directly regulate members of that community: it cannot control the actions of persons external to the community.[5] The first order of business for any community is to therefore define its membership. A community is by definition a group of people who

---

4   Admittedly a large proportion of this information is communicated through the market by the fashion industry itself, so you may wish to define fashion as a hybrid competition/community control model.
5   Although the values of a community may indirectly affect members of other communities should that external community recognise the community values of the first community. This will be discussed further below.

share a common experience or interest.[6] But simply to define a community thus does not capture completely the boundaries of a community or its ability to regulate its members. There are in fact two types of community: those we are members of due to external aspects of our lives such as where we live (I am a member of the community of UK citizens because I live in the UK) or what we do (lawyers are members of a legal community; those who practice as lawyers are members of the Law Society, those who work as barristers are members of the Bar), and communities we actively choose to join such as arts societies, football supporters clubs, or simply 'membership' of the community in your local bar or coffee shop. The first type of community may be referred to as macro-communities: these are broad communities that impact extensively on an individual's day-to-day life. Such communities often have rigid community standards or norms that are designed, and enforced, to ensure that the values of the community are upheld. For example the Law Society regulates all aspects of its members' professional lives to ensure that trust and confidence in the legal profession and its members are upheld. The second type are micro-communities: these are narrow communities focused on a particular aspect of an individual's life, usually part of their social life. There tend to be lower barriers to entry and exit, and much greater mobility within micro-communities. This means that it is usually easier to join and to leave micro-communities, and the switching costs of so doing are usually much less than with macro-communities.[7] Once the community is identified it may regulate its members by adopting a set of community values. In the informal these values may be termed as standards: the level of behaviour that is expected from the members of the community. In the formal these values may be termed as norms: the level of behaviour that is required of members of the community and which, if not met, will lead to the expectation of rebuke or stronger response up to and including exclusion from the community. These values, whether they form standards or norms, should develop organically from the bottom up. It is not necessary that they reflect the values of all the members of the community, but they should reflect the values of at least the majority of the community. Values that are imposed upon community members, such as values imposed by a ruling committee, are not community standards or norms; rather they are an exercise of power and as such are a form of hierarchical regulation. Traditionally, it is the relationship between community values and hierarchy that has formed the focus of most discourse on community regulation.

---

6   The OED definition of a community reflects this.
7   Switching costs are costs incurred in changing suppliers or marketplaces (or in this case community). Examples of switching costs include the effort needed to inform friends and relatives about a new telephone number after an operator switch, costs related to learning how to use the interface of a new software package and costs in terms of time lost due to organisation and legal paperwork required to move home.

Communities, hierarchy and control

When dealing with communities, in particular macro-communities, it can be difficult to differentiate between established social norms and hierarchically imposed rules. The confusion is caused because members of macro-communities are usually part of the community by default rather than by choice. This is due, at least in part, to high switching costs exerted upon anyone who wishes to leave a macro-community, making it very difficult for individuals to remove oneself from membership of a macro-community. We can see this by referring to one of my earlier examples of macro-communities: the community of UK citizens. The community of UK citizens is a large macro-community, whose disparate membership is made up of individuals linked together by only physical proximity and who are also members of a variety of micro-communities[8] and smaller macro-communities.[9] In terms of order it is probably one of the highest-order macro-communities, it being part only of international communities such as the community of European citizens or the community of citizens of the world.[10] As a member of the UK community, I am, by default, regulated by its community standards and norms.[11] If I wish to insulate myself from regulation by these standards and norms my only available choices are to remove myself from the observation or interaction of the community. This means either physically relocating myself outside the UK, or withdrawing myself from community interaction. Both these options though carry high switching and opportunity costs. In the former, there are the issues of leaving behind family and friends, finding new employment, paying for removal expenses, relocating family members (including arranging for schooling) and many more. With the latter it requires one to take on an extremely withdrawn lifestyle where social interaction is minimised at all costs, this would effect one's employment and one's social life, as well as one's immediate physical environment. These existing high

8 Such as members of the Automobile Association, Supporters of Aberdeen Football Club or Friends of the Old Vic Theatre.
9 Such as professional communities including the Law Society and the Institute of Chartered Accountants for England and Wales, and smaller geographical macro-communities such as residents of the London Borough of Westminster or citizens of Scotland, England, Wales or Northern Ireland.
10 In truth, when we reach a macro-community of such order we mean the genus of humanity rather than a community. This is because there are remarkably few experiences in common once the community becomes this large. That being said, some basic shared experiences that cause us to speak out against hunger, torture or genocide suggest that it may be possible to discuss the community of citizens of the world. Such a community is though probably too diverse to effectively enforce its norms unless they are enshrined and supported by hierarchical organisations such as the European Union or the United Nations.
11 These (probably) include: respect for others; the protection of children and other vulnerable groups; decency; consideration and proper comportment (e.g. one will not drink to excess during the day).

barriers to exit mean that for the individual member of the community, the regulatory controls exercised by the community of citizens of the UK; are every bit as difficult to evade as the laws of the UK; in fact, given the large network of community detectors evading detection is arguably more difficult.[12] As both hierarchical rules – such as laws – and community value sets – such as norms and standards – exert external controls upon the actions of the individual, and as both use an external network of agents to detect breaches of these rules or standards – breaches which if detected carry the threat of sanction or punishment – it is therefore not difficult to see why the two are easily confused, especially when dealing with macro-communities where membership of the community is also dictated by external factors. Attempting to map the relationship between the community, norms and the exercise of hierarchical power has therefore been of particular interest to sociologists and legal philosophers. In particular, we find this complex relationship at the heart of social contract theory, and the work of the three great social contract theorists – Thomas Hobbes, John Locke and Jean-Jacques Rousseau – help explain how we have come to model this relationship, which is at the heart of most modern regulatory and political theory.

Social contracts and Hobbes, Locke and Rousseau

Social contract theory is almost as old as philosophy itself. It is the view that a persons' moral and/or political obligations are dependent upon a contract or agreement between the members of a community or society, a contract which binds us together to form society. An early variety of social contract theory was used by Socrates to explain to Crito why he must remain in prison and accept the death penalty. The discussion between Socrates and Crito was recorded in Plato's classic text, 'Five Dialogues'.[13] According to Plato's account of that meeting a key argument put forward by Socrates in favour of accepting the death penalty that had been handed down to him was as follows:

> Will they not say: 'You, Socrates, are breaking the covenants and agreements which you made with us at your leisure, not in any haste or under any compulsion or deception, but having had seventy years to think of them, during which time you were at liberty to leave the city, if we were

---

12 In truth you are unlikely to be prosecuted if you commit a criminal offence: first, you will need to be apprehended; second, the authorities will need to have requisite evidence to proceed; and third, a jury or judge will have to be convinced of your guilt beyond all reasonable doubt. Social condemnation or censure is much more likely should you commit an antisocial act in public, despite the apparent drop in civic responsibility, which appears to have occurred in the UK in recent years.

13 Plato, *Five Dialogues*, Grube, G (trans), 2002, Indianapolis: Hackett Publishing.

not to your mind, or if our covenants appeared to you to be unfair. You had your choice, and might have gone either to Lacedaemon or Crete, which you often praise for their good government, or to some other Hellenic or foreign State. Whereas you, above all other Athenians, seemed to be so fond of the State, or, in other words, of us her laws (for who would like a State that has no laws?), that you never stirred out of her: the halt, the blind, the maimed, were not more stationary in her than you were. And now you run away and forsake your agreements. Not so, Socrates, if you will take our advice; do not make yourself ridiculous by escaping out of the city.'

Socrates here clearly expresses an early incarnation of social contract theory. The reasons he gives for his refusal to flee Athens are many. Central to his discussion with Crito is the concept of the Athenian State as his family: his educator, protector and guide. Crucially, Socrates saw that he had made an agreement or covenant with the State that in return for these benefits he would agree to abide by the rules and principles of the State. It is this bargain, a bargain we can in the terms described by Socrates accept, reject or seek to renegotiate, which is at the heart of the classical social contract theorists Hobbes, Locke and Rousseau.

Before looking at the work of the classical social contract theorists though, contemporary scholars of social contract theory should be aware of the rather different political and social values of seventeenth- and eighteenth-century England and France: the classical social contract theorists lived in a dramatically different world from that of Socrates, and from that of today. By the time Thomas Hobbes was born in the town of Malmesbury, Wiltshire in 1588, Athenian democracy had been dead for over 1,800 years.[14] The world that Hobbes was born into was one where there appeared to be a clear divide between those who ruled and those who were subjects to be ruled. Kings and Queens ruled by 'divine right', a political doctrine, which stated that the monarch drew his or her power to rule their subjects from the will of God, not from the will or agreement of their subjects. This meant that no matter how iniquitous or incompetent the monarch was, there was no right for them to be overthrown as to do so would directly interfere with the will of God. In 1642, the concept of the divine right was thrown into turmoil. The current monarch, King Charles, had, since 1638, been fighting a war with his subjects in Scotland, who had rebelled against his attempts to bring the practices of the Presbyterian Church of Scotland into line with the practices of the

---

14 The author acknowledges that Athens remained an important part of the Roman Empire until around AD 500 and that pockets of democracy had continued to thrive throughout medieval Europe, including early medieval Ireland, the Republic of Venice and the Nobles Democracy and *pacta conventa* of Poland.

Episcopalian Anglican Church. This short war, known as the Bishops' Wars, was a disaster for Charles: he lost a major battle at Newburn in 1640 and quickly lost the counties of Northumberland and Durham to Scots forces, and in October 1640, Charles was forced to sign the Treaty of Ripon. The Treaty stipulated that Northumberland and County Durham were to be ceded to the Scots as an interim measure, left Newcastle in the control of the Scots, and ordered that Charles was to pay the Scots £850 a day to maintain their armies there.[15] Charles was impoverished by the terms of the Treaty, a state of affairs that led him to recall Parliament to a sitting now known as the Long Parliament. Parliament, though, rather than issuing Charles the funds he needed to rejoin battle, instead sought to question his motives. Alarmed at this turn of events, Charles attempted to have five Members of Parliament arrested on charges of treason.[16] This measure failed though, and the five went into hiding. As the relationship between the King and Parliament continued to dissolve, Charles left London and gathered his troops at Nottingham. In August 1642, Charles raised his standard at Nottingham, signalling the start of the English Civil War.[17] Ultimately, the Parliamentarians defeated the Royalist armies and on 30 January 1649, King Charles I was beheaded at the Palace of Whitehall and for the first and only time in England's history it became a commonwealth under the protectorate of Oliver Cromwell. Against such a background even Socrates would have found it difficult to conceptualise the relationship between the State and its people.

Despite these upheavals, Thomas Hobbes opened a discourse into this relationship in his classic text *Leviathan*,[18] a discourse, which over the course of the next century was to be developed by John Locke and Jean-Jacques Rousseau. In a view perhaps influenced by the events of the previous 10 years, Hobbes believed that without government, men were doomed to live in the state of nature, a state in which violence and fear were prevalent. In the state of nature there are three constants: (1) individuals will violently compete to secure the basic necessities of life and perhaps to make other material gains; (2) individuals will challenge others and fight out of fear so as to ensure personal safety; and (3) individuals will seek reputation 'glory', both for its own sake and for its protective effects.[19] The effect of these is that in the state of nature each individual is forever in fear for their life, as alone, they are

---

15 For a full discussion of the Bishops' Wars see Fissel, M, *The Bishops' Wars: Charles I's Campaign Against Scotland, 1638–40*, 1994, Cambridge: CUP.
16 The five were John Hampden, John Pym, Arthur Haselrig, Denzil Holles and William Strode.
17 It is not the intent of this book to discuss the details of the Civil War, just to record its effects on Thomas Hobbes. Those interested in the detail of the War should read Young, P and Holmes, R, *The English Civil War: A Military History of Three Civil Wars, 1642–51*, 2000, Ware: Wordsworth.
18 Hobbes, T, *Leviathan*, 1651, Penguin Classics Edition, Harmondsworth: Penguin, 1968.
19 Ibid, 1.13, pp 61–3.

quite defenceless and they may be attacked and killed or maimed at any time. As Hobbes notes in one of his most celebrated passages:

> [I]n such condition there is no place for industry, because the fruit thereof is uncertain: and consequently no culture of the earth; no navigation, nor use of the commodities that may be imported by sea; no commodious building; no instruments of moving and removing such things as require much force; no knowledge of the face of the earth; no account of time; no arts; no letters; no society; and which is worst of all, continual fear, and danger of violent death; and the life of man, solitary, poor, nasty, brutish, and short.[20]

Hobbes' response to this state of nature is to suggest a rudimentary form of social contract. He suggests that as individuals are unable to rely indefinitely on their individual powers in the effort to secure livelihood and contentment it is logical for human beings to join together to form a 'commonwealth'.[21] Hobbes believed the commonwealth to be 'one person, of whose acts a great multitude, by mutual covenants one with another, have made themselves every one the author, to the end he may use the strength and means of them all as he shall think expedient, for their peace and common defence'.[22] Thus, the commonwealth as a whole embodies a network of associated contracts and provides for the highest form of social organisation. On Hobbes' view, the formation of the commonwealth creates a new, artificial, person (the Leviathan) to whom all responsibility for social order and public welfare is entrusted. Of course, someone must make decisions on behalf of this new whole, and that person will be the sovereign. Thus the commonwealth-creating covenant is not in essence a relationship between subjects and their sovereign, rather it is a social contract managing the relationship among subjects, all of whom agree to divest themselves of their native powers in order to secure the benefits of orderly government by obeying the dictates of the sovereign authority.[23] This is why the minority, who might prefer a different sovereign authority, have no complaint, as on Hobbes' view even though they have no respect for this particular sovereign, they are still bound by their contract with fellow subjects to be governed by a single authority. The sovereign is thus the institutional embodiment of orderly government.

Today, Hobbes' definition of the commonwealth, centred upon a sovereign ruler, may look rather unusual, but for Hobbes, writing in a time of great turmoil when England had gone through 10 years of civil war and had seen power transferred from a monarch appointed in accordance with the Divine

20 See Hobbes, above, fn 18, 1.13, p 62.
21 The choice of term being no doubt influenced by the current political settlement in England.
22 Ibid, 2.17, p 88.   23 Ibid, 2.18, pp 88–9.

Right of Kings, to a dictatorial Lord High Protector, it seemed like the natural order. For Hobbes, who had been maths tutor to the future King Charles
II, and who had fled to France in 1640, the power of the sovereign was at the
heart of social order. Hobbes' view of the community was merely as a shelter
from anarchism: decisions were made by the sovereign, whether he be a hereditary monarch, a legislature or an assembly of all citizens, or an individual.
Hobbes himself suggests the best monarch is a single natural person who can
choose advisors and rule consistently without fear of internal conflicts.[24]
Thus a Hobbesian view of the social contract is one where the subjects of the
sovereign agree to be subject to his common authority. Submission to the
sovereign is absolutely decisive. The structure provided by orderly government, according to Hobbes, enhances rather than restricts individual liberty.
Ultimately therefore, the Hobbesian covenants are about validating the exercise of hierarchical power rather than social norms and standards. For
Hobbes, the only role of the individual is to succumb to the will of the
monarch. He does not address what the role of the individual is as a member
of the macro-community and therefore, ultimately, reading Hobbes is disap-
pointing for the scholar of regulatory theory. Hobbes remains throughout a
classical legal scholar; he assumes that power and control are ultimately one
and the same. If we hope to be able to distinguish between the controls
exercised over individuals through their membership of the macro-
community and controls exercised over individuals because they are subject
to a hierarchical control system, we must look elsewhere.

John Locke is the second key social contract theorist of the golden age.
Whereas Hobbes saw man as being evil, Locke views man in a much more
optimistic light. They both agree that all men are equal according to natural
law, however, their ideas of natural law differ greatly. Whereas Hobbes saw
natural law as a state of war in which 'every man is a enemy to every man',
Locke sees natural law as a state of equality and freedom.[25] Locke therefore
believes that government is necessary in order to preserve natural law, while
Hobbes sees government as necessary in order to control natural law. Since
natural law is a positive state for Locke, it is the role of government to
preserve and enhance natural law, rather than seek to control it.[26] For Locke,
social orders are of key importance. Unlike Hobbes, Locke sees social constructs such as the family and community as being natural orders and this
influences his views on the relationship between community controls and
hierarchical controls. The first instance of social organisation, on Locke's
view, is the development of the family, a voluntary association designed to

24 Thus Hobbes' personal preference is for a hereditary monarch, Ibid 2.19, p 98.
25 Locke, J, *Second Treatise of Government*, 1690, Dover Edition, Minneola, NY: Dover
     Publications, 2002, 2.4.
26 Ibid, 9.123.

secure the propagation of the human species through successive generations.[27] Each family will develop its own property interests, and in the state
of nature each has the right to enforce the natural law in defence of these
property interests. For Locke though, the distinction between the state of
nature and civil society is not in the removal of a state of fear, but rather in
the voluntary submission to the standards and values of the community at
large. Locke believed the formation of a civil society required that all individuals voluntarily surrender their right of defence to the community at large.
By declaring and enforcing fixed rules for conduct the commonwealth thus
serves as 'umpire' in the adjudication of property disputes among those who
choose to be governed in this way.[28] Securing social order through the formation of any government invariably requires the direct consent of those who
are to be governed.[29] Each and every individual must concur in the original
agreement to form such a government, but it would be enormously difficult to
achieve unanimous consent with respect to the particular laws it promulgates.
So, in practice, Locke supposed that the will expressed by the majority must
be accepted as determinative over the conduct of each individual citizen who
consents to be governed at all.[30] By introducing an element of democracy,
and democratic accountability into the social contract, Locke provides a key
element in the explanation of the relationship between macro-community
values and the exercise of hierarchical controls. Unfortunately, as with the
earlier work of Hobbes, the State he discusses is too remote from ours to
provide a full explanation of the modern social contract. What Locke provides is a move away from the assumption that life without the direction of
a sovereign ruler is destined to be violent and bloody. Locke recognises the
key role of the family and, in particular, the social relationship of parent and
child, a complex relationship – part hierarchical, part social.[31] Locke has a
more sophisticated understanding of the relationship between the State and
its citizens, recognizing that like the family relationship, it is a relationship
based in social interaction as well as the application of power.

Both Locke and Hobbes brought social contract theory back into consideration. For too long communities had been ruled by absolute rulers with
almost limitless power. The concept of democracy in its true form had long
been lost and social norms and standards had been subjudicated by the
power of the sovereign. Although Hobbes and Locke, and in particular
Hobbes, had ultimately used the concept of the social contract or covenant to
explain the authority of the State or sovereign to govern, they had also begun
a key process of evaluating the power of the individual. Hobbes had assumed

---

27 See Locke, above, fn 25, 5.78.   28 Ibid, 7.87–7.89.
29 Ibid, 8.95.   30 Ibid, 8.97–8.98.
31 Locke notes: 'The power, then, that parents have over their children, arises from that duty
    which is incumbent on them, to take care of their off-spring, during the imperfect state of
    childhood.', 6.58.

that governance by a sovereign as part of the commonwealth was necessary, as without such leadership, life would be violent and short. Locke, making no such assumption, suggested that individuals agree to be governed to provide a fair, impartial and standardised system of law and punishment, but that this political state is justified only by the consent of the people who presume that the State will protect their natural rights of life, liberty and property. The people freely agree through the social contract to abide by the laws that they or their elected representatives enact: in this way they are bound by law. The next logical stage of development of social contract theory was to examine this contract between the State and its citizens. This was taken on by Jean-Jacques Rousseau, who was born in Geneva on 28 June 1712, some seven years after the death of John Locke.

Rousseau has two distinct social contract theories. The first is an account of the moral and political evolution of humans from a state of nature to modern society. This is his naturalised account of the social contract, an account that troubled Rousseau and that was developed in his essay, *Discourse on Inequality*,[32] commonly referred to as the Second Discourse. Rousseau wrote his Second Discourse in response to an essay contest sponsored by the Academy of Dijon. In it he describes the historical process by which man began in a state of nature and over time progressed into civil society. According to Rousseau, the state of nature was a peaceful and idealistic time. People lived solitary, uncomplicated lives. Their few needs were easily satisfied by nature. Because of the abundance of nature and the small size of the population, competition was nonexistent, and persons rarely even saw one another, much less had reason for conflict or fear. Furthermore, these simple, morally pure persons were naturally endowed with the capacity for pity and therefore were not inclined to bring harm to one another. As time passed, though, humanity faced certain changes. As the overall population increased, the means by which people could satisfy their individual needs changed. People slowly began to live together in small families, and then in small communities. Divisions of labour were introduced, both within and between families, and discoveries and inventions made life easier, giving rise to leisure time. Leisure time inevitably led people to make comparisons between themselves and others, resulting in public values, leading to shame and envy, pride and contempt. Most importantly however, according to Rousseau, was the invention of private property which constituted the pivotal moment in humanity's evolution out of a simple, pure state into one characterised by greed, competition, vanity, inequality and vice. For Rousseau the invention of property constitutes humanity's fall from the state of nature.[33]

32 1754, Oxford World Classics Edition, Philip, F (trans), 1999, Oxford: OUP.
33 Rousseau notes: 'This was the epoch of a first revolution, which established and distinguished families, and introduced a kind of property, in itself the source of a thousand quarrels and conflicts.' Ibid Part II.

Having introduced private property, initial conditions of inequality became more pronounced. Some have property, while others are forced to work for them, and the development of social classes begins. Eventually, those who have property notice that it would be in their interests to create a government that would protect private property from those who do not have it, but who can see that they might be able to acquire it by force. So, governments are established, through a contract, which purports to guarantee equality and protection for all, even though its true purpose is to fossilise the very inequalities that private property has produced. In other words, the contract, which claims to be in the interests of everyone equally, is really in the interests of the few who have become stronger and richer as a result of the developments of private property. This is the naturalised social contract, which Rousseau viewed as responsible for the conflict and competition from which modern society suffers.

The second theory is his normative, or idealised, theory of the social contract, and is meant to provide the means by which to alleviate the problems that modern society has created for us. The normative social contract, argued for by Rousseau in *The Social Contract*,[34] is meant to respond to this sorry state of affairs and to remedy the social and moral ills that have been produced by the development of society. The distinction between history and justification, between the factual situation of mankind and how it ought to live together, is of the utmost importance to Rousseau. While we ought not to ignore history, nor ignore the causes of the problems we face, Rousseau believes we must resolve those problems through our capacity to choose how we ought to live. *The Social Contract* begins with the most frequently quoted line from Rousseau: 'Man was born free, and he is everywhere in chains.' This claim is the conceptual bridge between the descriptive work of the Second Discourse, and the prescriptive work that is to come. Humans are essentially free, and were free in the state of nature, but the progress of civilisation has substituted subservience to others for that freedom. Since a return to the state of nature is neither feasible nor desirable, the purpose of politics is to restore freedom to us, thereby reconciling who we truly and essentially are with how we live together. This is the fundamental philosophical problem that *The Social Contract* seeks to address: how can we be free and live together? Or, put another way, how can we live together without succumbing to the force and coercion of others? We can do so, Rousseau maintains, by submitting our individual, particular wills to the collective or general will, created through agreement with other free and equal persons. Like Hobbes and Locke before him, Rousseau believes all men are made by nature to be equals, therefore no one has a natural, or divine, right to govern others, and therefore the only justified authority is the authority that is generated out of agreements or

34 1762, *Penguin Books Great Ideas*, Cranston, M (trans), 2004, London: Penguin.

covenants. The most basic covenant – the social pact – is the agreement to come together and form a people, a collective, which by definition is more than, and different from, a mere aggregation of individual interests and wills. This act, where individual persons become a people is 'the real foundation of society'.[35] Through the collective renunciation of the individual rights and freedom that one has in the state of nature, and the transfer of these rights to the collective body, a new person is formed. The sovereign is thus formed when free and equal persons come together and agree to create themselves anew as a single body directed to the good of all considered together. So, just as individual wills are directed towards individual interests, the general will, once formed, is directed towards the common good, understood and agreed to collectively. Included in this version of the social contract is the idea of reciprocated duties: the sovereign is committed to the good of the individuals who constitute it, and each individual is likewise committed to the good of the whole. Given this, individuals cannot be given liberty to decide whether it is in their own interests to fulfill their duties to the sovereign, while at the same time being allowed to reap the benefits of citizenship. They must be made to conform themselves to the general will, or in Rousseau's words, they must be 'forced to be free'.[36] For Rousseau, this implies an extremely strong and direct form of democracy. One cannot transfer one's will to another as one does in representative democracies. Rather, the general will depends on the coming together periodically of the entire democratic body, each and every citizen, to decide collectively, and with at least near unanimity, how to live together, that is what laws to enact. As it is constituted only by individual wills, these private, individual wills must assemble themselves regularly if the general will is to continue. One implication of this is that the strong form of democracy that is consistent with the general will is also only possible in relatively small communities. The people must be able to identify with one another, and at least know who each other are. They cannot live in a large area, too spread out to come together regularly, and they cannot live in such different geographic circumstances as to be unable to be united under common laws. Although the conditions for true democracy are stringent, they are also the only means by which we can, according to Rousseau, save ourselves, and regain the freedom to which we are naturally entitled. Rousseau's social contract theories together form a single, consistent view of our moral and political situation. We are endowed with freedom and equality by nature, but our nature has been corrupted by our contingent social history. We can overcome this corruption, however, by invoking our free will to reconstitute ourselves politically, along strongly democratic principles, which is good for us, both individually and collectively.

35 Book I, Ch 5.   36 Book I, Ch 7.

Some attempts have been made to apply social contract theory to contemporary thought, in particular through the works of the legal philosopher John Rawls[37] and the moral philosopher David Gauthier.[38] Both Rawls and Gauthier examined the social contract against abstract questions of morality and justice. There have been few attempts to apply the social contract theories of the classical scholars to contemporary social organisations. This may be because the theories of the classical scholars have been heavily critiqued as being too simplistic, or too dated by most contemporary legal philosophers.[39] The value of the social contract as a model is therefore heavily discounted. But, it may be argued that the value of the social contract is to be found in the very simplicity that modern scholars discount as an inherent weakness. In particular, it may be argued that a comparison of the works of Locke and Rousseau may explain the relationship between the community and the sovereign in a way that helps us clarify the distinction between the normative controls of the macro-community and the exercise of hierarchical power.[40] The distinction between Locke's covenant and Rousseau's social contract is to be found in the way power is exercised. Locke suggests that whereas the original covenant, granting authority to the sovereign or government, must be universally accepted by the members of that society, once the terms of the covenant are agreed, the will of the majority may be enforced over the views of the minority. This suggests representative government where the views of the majority are represented and enforced over the views of the minority, or to express the relationship differently in a Lockean covenant by agreement, the community agree to relinquish power to the sovereign or government, within the terms of the covenant. Thus the members of the community have agreed to be bound by the decisions of the rule-makers, provided they do not exceed the authority given to them by the community. This defines

37 In his 1971 book *A Theory of Justice* (Cambridge, MA: Belknap Press) Rawls suggested that a just social contract is that which we would agree upon if we did not know in advance where we ourselves would end up in the society that we are subscribing to. This condition of ignorance is known as the original position. In the original position, each person would not know their financial situation, race, creed, religion, or state of health. From behind this veil of ignorance, we can discern the form of a truly just society, since our judgment would not be clouded by knowledge of our own personal interests. Rawls later retreated from this position in his later work. Although he never retreated from the key arguments of *A Theory of Justice*, later in life he developed his theories of political liberalism, which softened his views somewhat. See Rawls, J, *Political Liberalism*, 1993, New York: Columbia UP.
38 Eg Gauthier, D, *Moral Dealing: Contract, Ethics, and Reason*, 1990, Ithaca, NY: Cornell UP.
39 Eg *Critical Essays on Hobbes, Locke and Rousseau: The Social Contract Theorists*, 1999, Lanham, MD: Rowman & Littlefield.
40 Although the position of sovereign is obviously a position of particular significance in any society, it may be argued that a similar line of analysis may be taken with any party in a position of power, and therefore any hierarchical relationship may be analysed in the following way.

hierarchical regulation: when someone in a position of authority over members of a community (either a macro-community or a micro-community) makes a ruling or promulgates a new rule, principle or law, they do so with the agreement of the community.[41] Importantly, this does not have to be the agreement of all the community – only a majority of the community. Equally, an abuse of their position will lead to an invalid or 'bad' law. This is not to say that such 'bad' laws are any less effective in regulating the immediate actions of others. The mugger with the gun controls the actions of his victim by the application of power, but he has done so without the agreement of the community. Once the community reasserts its control over those who make bad laws their actions laws, rules or principles may be rescinded, and thus the mugger will be suitably punished and the victim will have their property returned, if possible. Alternatively, the definition of community-based controls may be found in Rousseau's normative social contract. Rousseau believed that the general will of the community constituted the sovereign body and that this could not be divorced from members of the community. Thus in Rousseau's normative social contract the community retains power to itself as the sovereign or government. Decisions are made from within the body of the community for the collective benefit of all members of the community. Locke describes an exercise in hierarchical power within a macro-community: power given by covenant to a ruling body, hierarchy or sovereign, whereas Rousseau describes an exercise in community decision-making: power retained within the body of the community to make decisions for the benefit of the community. Thus identifying where the power to promulgate new principles, norms, rules or laws resides, provides an easy distinction between the exercise of hierarchical power and the exercise of community controls. With this knowledge it is easy to identify which rules and principles originate as community-based rules and which originate as hierarchical controls.

Online macro-communities: netiquette

The nature of the internet mitigates against the creation of macro-communities. We do not 'live' in the online environment; rather it is somewhere we visit or travel to for business or for pleasure. Cyberspace has few barriers to exit: to leave one need only 'switch off' and you are returned to the real world. Thus it is difficult to conceive of online equivalents of professional societies such as the Law Society or geographical communities, such as residents of Greater London. It should be noted that in making this claim I do not denigrate the feelings of inertia felt by committed online games players or eBay traders when time comes for them to vacate their online

41 This agreement may be tacit rather than explicit.

community and online life, but these individuals are not physically or pro-
fessionally tied as members of macro-communities are in the real world.
The nature of the network is such that there is probably no identifiable
macro-community in cyberspace, but one aspect of community regulation
may be discussed under the heading of the macro-community and this is the
existence and use of the 'netiquette' system across the network.

Although netiquette does not apply to all uses of the internet it is probably
the widest form of community regulation in cyberspace. Netiquette is a
catch-all term for the conventions applied on Usenet, in mailing lists, and in
other electronic fora such as message boards and discussion fora. The origins
of netiquette are another of the large body of internet phenomena that seem
to be traceable to a small group of computer scientists in the 1970s, with most
early netiquette rules appearing to be as much an operation of design control
as community control. Early netiquette rules such as 'Do not post off-topic
materials', 'Do not cross-post' and 'Do not quote in posts' were developed to
deal with the problem of restricted bandwidth: the underlying principle of all
three rules being 'Do not post anything which costs the system bandwidth
that has not been budgeted for'.[42] As bandwidth became less restrictive, rules
were developed to help bridge the gap between the physical world and the
virtual. The conveying of emotions was one of the first issues to be tackled.
Expressing anger became possible THROUGH THE USE OF CAPITAL
LETTERS TO DESIGNATE SHOUTING, but equally a rule of etiquette
developed saying capitals should not be used unless specifically for this reason. Other emotions such as happiness, anger or surprise as well as some
physical characteristics could be conveyed through the use of emoticons:
small three- or four-character text pictures used to indicate facial communication.[43] Another important aspect of internet communication that came to
the fore during the development of the online community in the 1970s and
1980s was how to establish one's identity in a non-physical environment.
Most internet users could not hope to have a home email account at this time,
and therefore one's identity became entwined with where one worked. It was
important to the community that they could identify the individual, rather
than simply the email account. Therefore, the third major development in
netiquette at this time became the signature block. The short-text signature

---

42 *A History of some Usenet Rules*, Usenet Information Center, at *www.usenetmonster.com/
infocenter/articles/usenet_netiquette.asp*.
43 Emoticons were first suggested in September 1982 by Scott E Fahlman as a method of
distinguishing jokes in postings. He suggested jokes be marked :-) while serious posts be
marked :-(. Over time emoticons have become commonplace and now are used to indicate
hundreds of different short messages. The Computer User High-Tech Dictionary currently
lists 239 different short text emoticons, including >>:-<< (Furious); :-Q~ (Smoking) and
@>—>— (A long-stemmed rose). The full list is at: *www.computeruser.com/resources/
dictionary/emoticons.html*.

block was a method of identifying the individual, as distinct from their employer. Signatures usually contained the individual's name, position and email address and would remain attached to their original message even if the message headers were removed. Always mindful of the overriding rule of netiquette of 'do not post anything which costs the system bandwidth that has not been budgeted for', it is considered good netiquette to keep your signatures short, usually under six lines, so that they do not take up excessive space in recipient's mailbox.

These early rules were part of the fabric of the internet community. As the community grew and the technology developed, the rules of netiquette evolved. In the early 1990s, though, the previous tradition of a flexible, unwritten code, changed. The development of the web had caused a massive increase in size in the internet community and most of those who took advantage of this development were network virgins, or in the terminology of the community, newbies. To alleviate the risk that through their unfamiliarity with the rules of netiquette, newbies would inadvertently cause offence or disruption, Sally Hambridge of Intel recorded the community's informal rules and published them as RFC 1855 *Netiquette Guidelines*.[44] By publishing RFC 1855 the community elders, through Hambridge, crystallised their informal standards into a set of formal rules. These formal rules took two clear and distinct forms: a set of soft regulations setting out the community standards with respect to acceptable forms of language, decency, respect for the property and privacy of others, and a set of hard rules covering technical standards such as the length of messages, form of messages, how to forward and reply to messages and the storage of information. The former are designed to act as simple reminders and guidelines, even platitudes. They contain guidance such as: 'Remember your recipient is a human being whose culture, language and humor have different points of reference from your own. Date formats, measurements and idioms may not travel well. Be especially careful with sarcasm'[45] and 'Wait overnight to send emotional responses to messages';[46] while the latter are prescriptive rules such as 'Limit line length to fewer than 65 characters and end a line with a carriage return',[47] and 'Use two carriage returns to indicate that you are done and that the other person may start typing'.[48] The rules of netiquette as recorded in RFC 1855 thus reflect the standards of a community confident in technology, but less confident in dealing with other people. Around the same time the first published guide to netiquette appeared. In her book, *Netiquette*,[49] Virginia Shea explained more fully for newbies how to venture into cyberspace without inadvertently causing offence. This guide offered

---

44 Hambridge, S, *Netiquette Guidelines*, 1995, RFC 1855, available at *www.ietf.org/rfc/rfc1855.txt*.
45 RFC 1855, 2.1.1.   46 Ibid.   47 Ibid.   48 Ibid, 2.1.2.
49 Shea, V, *Netiquette*, 1994, San Rafael, CA: Albion Books.

practical advice such as 'brush up on your spelling and grammar: you will be judged by the quality of your writing'[50] and 'never send a file [by e-mail] when a simple note would do'.[51] It may have been expected that with a constant stream of newbies joining the network in the 1990s and into the new millennium that interest in netiquette would remain strong. Unfortunately it appears that the concept of a single body of netiquette is fast becoming an anachronism. Hambridge's guide has never been updated and is now vastly out of date. Similarly, Shea's book never went into a second edition while publishers seem resistant to producing any further guides on netiquette.[52] Why is this? It appears that although we may imagine netiquette to be a single set of values for the internet community as a whole, it ceased to be so after the explosive growth of the network seen in the early 1990s. During this period of rapid expansion a variety of micro-communities developed. Some of these new communities adopted netiquette in full, while some adopted a modified version of netiquette. Most, though, were communities developed to take advantage of the opportunities offered by the new consumer-friendly technology offered by the web. As such, the 'old' rules developed in menu-driven systems such as Usenet, FTP and Gopher, did not apply. These communities therefore developed their new standards and principles drawing upon some of the established rules of netiquette, but adapted to suit the new environment. Thus today it is the case that although some standards of netiquette are still widely applied,[53] most internet users would find it difficult to identify the etiquette of cyberspace.[54] This is a simple reflection of the fact that the network has grown and is now too large to effectively leverage its macro-community values. In the modern network we are more immediately affected by the rules of the micro-communities of which we are members. The community of all internet users is now too big to be an effective community with coherent values and principles. We no longer travel to cyberspace: we now travel to destinations within cyberspace such as EverQuest, Project Entropia, eBay, or The Well. It is to these communities we must look if we are to identify the root of community regulation in cyberspace.

50 See Shea, above, fn 49, p 41.   51 Ibid, p 53.

52 Amazon.com lists only four books on netiquette, including Shea's, while listing 1,919 books on etiquette.

53 Standards such as not typing in caps, the correct use of emoticons, apologies for cross-posting and control of spam are still widely applied.

54 An ICM Survey carried out in January 2001 found that two-thirds of the 18–24-year olds questioned do not worry about punctuation, grammar or style when writing email messages and about 16 per cent sign every email with love and kisses, even when addressing their boss. This result caused the BBC to note that 'Good manners in e-mail are rarer than sensible hats at Ascot'. See: *http://news.bbc.co.uk/1/hi/sci/tech/1234233.stm*.

Online micro-communities

As in the real world there are an extensive variety of online micro-communities. These communities have standards and norms that are designed to reflect the aims and objectives of that community. Micro-communities can be categorised and identified using several typologies. Early attempts to categorise such communities often followed a technological taxonomy. This may be seen in Marc Smith and Peter Kollock's classic paper *Communities in Cyberspace*,[55] where the authors' note that '[e]ach online communication system structures interaction in a particular way, in some cases with dramatic effect on the types of social organizations that emerge from people using them'.[56] This leads the authors to examine communities built around email and discussion lists, Usenet and bulletin board systems (BBSs), chat, MUDs (multi-user dungeons), the web, and graphical environments. This topology is reflected throughout the remainder of their book, particularly in the chapters on identity in cyberspace by Judith Donath[57] and Elizabeth Reid's chapter on hierarchy and power,[58] which focus almost exclusively on Usenet and MUD's, respectively.[59] Such attempts to model complex community constructs through simple concepts such as disintermediation, interfaces, permissions, role-playing and pseudonymity were soon criticised. One strong critic of this approach was Manuel Castells, who in his book *The Internet Galaxy* noted that such an approach led to a 'rather sterile debate'.[60] Castells criticised the widely held belief that the internet leads to a breakdown in social communication and family life and an increasing abandonment of face-to-face interaction in real settings.[61] He notes that the user's experiences of the internet are 'overwhelmingly instrumental and closely connected to the work, family and everyday life of Internet users' and that '[r]ole-playing and identity building as the basis of on-line interaction are a tiny proportion of Internet-based sociability'.[62] Castells believes that examinations of online micro-communities should take a more catholic approach and that the traditional preoccupation with gaming communities, MUDs and Usenet groups is not terribly instructive. Instead he notes that the internet allows us to experience a different kind of community support: support for what he calls 'networked individualism'.[63] Network individualism is defined by Castells in two ways: first, the creation of new weak ties between people who share

55 Smith, M and Kollock, P, 'Communities in Cyberspace' in Smith, M and Kollock, P (eds), *Communities in Cyberspace*, 1999, London: Routledge.

56 Ibid, pp 4–5.

57 Donath, J, 'Identity and Deception in the Virtual Community' in Smith, M and Kollock, P (eds), *Communities in Cyberspace*, 1999, London: Routledge.

58 Reid, E, 'Hierarchy and Power: Social Control in Cyberspace' in Smith, M and Kollock, P (eds), *Communities in Cyberspace*, 1999, London: Routledge.

59 Ibid, pp 5–8.   60 Castells, M, *The Internet Galaxy*, 2001, Oxford: OUP at p 117.

61 Ibid, pp 117–25.   62 Ibid, p 118.   63 Ibid, pp 129–33.

some characteristics in common, an example of such a community being SeniorNet, a network designed to bring together elderly people for support and the exchange of information; and second, through technologies such as email and chat, the network provides for the maintenance of existing strong ties such as family ties, which have been stretched by geographical relocation. Thus Castells sees the internet as primarily a tool of communication that can be used to form and strengthen bonds between persons who have interests or experiences in common. Communities in cyberspace come together as an extension of communities in real space, with the power of the network being in its ability to shrink distances and time. Should you want to meet with others who share your interests in, for example, Ferrari cars, there is no need to travel to meet with others at a common time and place convenient to all. You may 'dip in' to a discussion at ferrarichat.com or thescuderia.net at any time and from the comfort of your own home. This leads Castells to note: 'the most important role of the Internet in structuring social relationships is its contribution to the new pattern of sociability based on individualism'.[64] The work of Castells, and others such as AJ Kim[65] and Barry Wellman,[66] caused sociologists and information systems researchers to formulate alternative taxonomies of online micro-communities, which were not dependent upon the technological foundations of the community.

To give researchers a greater understanding of the dynamics of online micro-communities, two alternative taxonomies were developed in the late 1990s in an attempt to more fully map the complex systems and relationships that occur in these communities. The first was founded in social network analysis and was represented most clearly by Barry Wellman's social network theory and Manuel Castells' networked individualism. This examined and categorised communities by reference to the social networks of the community members rather than the space the community inhabited. Wellman, in the chapter he co-wrote with Milena Gulia in *Communities in Cyberspace*,[67] promoted the value of social network analysis by noting that 'computer mediated communication accelerates the ways in which people operate at the centers of partial, personal, communities switching rapidly and frequently between groups of ties',[68] while Castells notes that with the development of strong online networks, 'individuals build their networks on-line and off-line, on the basis of their interests, values affinities and projects . . . online social interaction [will play] an increasing role in social organization as a whole'.[69] While the social network taxonomy of online communities is

---

64 See Castells, above, fn 60, p 130.
65 Kim, A, *Community Building on the Web*, 2000, Berkeley: Peachpit Press.
66 Eg Wellman, B and Gulia, M, 'Virtual Communities as Communities' in Smith, M and Kollock, P (eds), *Communities in Cyberspace*, 1999, London: Routledge.
67 Ibid.   68 Ibid, p 188.   69 Castells, above, fn 60, p 131.

valuable in understanding the fluid structure of these communities, and helps to explain the relationship between individual members and between the community and the member, it does not assist the regulatory theorist in mapping how, and through what means, community controls are exercised. It is a sociological model of online communities, not a socio-regulatory model and as such is of limited value to the modelling of regulatory controls. The second was to study the dynamics of online communities. In her book *Community-Building on the Web*,[70] Amy Jo Kim argued that online and offline communities are substantively the same, and that therefore traditional methods of modelling community dynamics may be used in modelling online communities.[71] By focusing on the dynamics of the community, one can identify the aspects of the community, which attract members to the community and which incentivise those members to return frequently to the community and to become involved in the community. Kim's dynamic modelling topology was further developed by Robert McArthur and Peter Bruza in their paper *The ABC's of Online Community*.[72] In this paper the authors propose that an online community has three basic characteristics: (1) people are essential and integral; three or more people may form a community; (2) computer systems and adequate access to them are essential; and (3) people have some language in common.[73] These are referred to by the authors as the 'core attributes' or 'conditions' of all online communities, but these attributes only facilitate the creation of a community, not its focus or its topology. McArthur and Bruza therefore go on to suggest that there are four key components that may be used to identify or model an online community: (1) Purpose;[74] (2) Commitment;[75] (3) Context;[76] and (4) Infrastructure.[77] The authors argue that these four components are necessary for a sustainable community to manifest itself in the online environment, and that without the 'glue' supplied by all of these components a community will

---

70 Kim, see above, fn 65.
71 At p 1, Kim notes: 'How is a Web community different than one in the real world? In terms of their social dynamics, physical and virtual communities are much the same. Both involve developing a web of relationships among people who have something meaningful in common, such as a beloved hobby, a life-altering illness, a political cause, a religious conviction, a professional relationship, or even simply a neighborhood or town. So in one sense, a Web community is simply a community that happens to exist online, rather than in the physical world.'
72 McArthur, R and Bruza, P, 'The ABC's of Online Community' in *Proceedings of First Asia Pacific Conference on Web Intelligence*, 2001, London: Springer-Verlag.
73 Ibid, p 143.
74 Purpose is the shared common goal, interest or theme of the community.
75 Commitment is the repetitive and active participation of the members of a community in the pursuit of its purpose.
76 Context is the attributes of the community which remain more or less fixed or stable, thereby providing the secure foundations on which the community can be built.
77 Infrastructure refers to the physical infrastructure used to support the community.

ail and die. Such dynamic modelling of online communities is valuable, and is particularly useful in modelling patterns of regulatory control, especially for systems designers, and for regulatory theorists developing rules-based and design-based control systems. By applying the dynamic modelling system, I believe that we can identify the foundations for the development of community-based regulatory systems within online micro-communities. In particular, as noted by McArthur and Bruza, purpose is particularly important for any community and to this end it is proposed that when examining the foundations of community-based regulation within modern online micro-communities we apply a taxonomy based upon the purpose of that community.

A modern taxonomy of online micro-communities

Using the purpose of the community as our primary reference point it is suggested that six classes of online micro-community, each with distinct primary purposes, may be identified.

Classification of micro-communities based upon their primary purpose is particularly valuable in the online environment. As there is no physical glue to hold people together in cyberspace, what draws individuals together to form a community is a shared set of values and goals. Further, given the almost infinite variety of online communities that currently exist, communities that can accommodate such diverse views as those held by Ku Klux

Table 5.1 Classification of communities by purpose

Class	Primary purpose	Examples
Commercial communities	Market functions, trade, payment, transactions, trust	EBay, QXL, yourcarboot.com, Ubid, Pirate Bay, etc.
Online/offline communities	Online discussion of offline subjects	britneyboards, Outpost Gallifrey, talkclassical.com, etc.
Gaming communities	Gaming progress, rewards, respect of other gamers	Everquest, Runescape, etc.
Cafe communities	Discussion and chat	The Well, ezBoard, MySpace, etc.
Knowledge communities	Help and advice, distribution of information and knowledge	Tripadvisor, Review Centre, Medical databases such as mdchoice.com, etc.
Creative communities	The creation of co-operative products and services	Wiki, (in particular wikipedia), GNU/Linux, etc.

Klan,[78] the Nation of Islam,[79] the National Alliance[80] and the Nizkor pro-ject,[81] there is almost certainly a community in place that meets closely the values of any individual. Thus cybercommunities may be much narrower communities than their real-world counterparts, with much more focused aims. An individual is likely to join several communities, each of which reflects some aspect of his or her personality, and their choices as to which communities they will join will be heavily influenced by whether or not that community can meet their goals. This is why a goals- or purpose- based classification of online communities is so valuable.

In the above classification (see Table 5.1), each of the six community types has a distinctive primary purpose. 'Commercial' communities such as eBay or Ubid are designed for those who wish to trade or exchange goods and/or services through the community. Such communities seek to streamline and simplify market functions and have developed systems to allow for trade and payment and have developed trust-based systems between the buyer and the seller. 'Online/offline' communities like The Leaky Cauldron (Harry Potter discussion) or Outpost Gallifrey (Doctor Who discussion) are communities that are driven by events that take place offline, but which use the network to facilitate meetings and discussions over time and distance. These com-munities tend to develop complex norms and structures designed to manage the divide between offline events and the online discussion of them. A key aspect of most of these communities is the use of the 'spoiler' notice when discussing items or events that not all members may yet have experienced. Thus although members of the Leaky Cauldron may all wish to discuss the events of the latest Harry Potter book, it is recognised that not all members read at the same pace, thus discussion of the latest developments must not be carried out in general discussion boards and is restricted to areas of the site where members may expect it to be discussed. 'Gaming' communities are among the oldest online communities. Gaming communities find their origins in the MUDs (multi-user dungeons) of the early 1980s and unlike the other communities, they are rooted in escaping from the real world rather than in supplementing the member's offline experiences. Gaming communities are exceedingly diverse and range from complex fantasy universes where players play the role of fantastical characters to simple online gambling communities such as PartyPoker.com. All gaming communities share though a common purpose of progress, reward and earning the respect of other community members for their gaming prowess, which sets them apart from other online

---

78 *www.kkk.com/*.   79 *www.noi.org/*.

80 The National Alliance is an extreme right-wing political organisation based in the US, which denies much of the WWII holocaust, *www.natall.com/*.

81 The Nizkor project is a web service, 'dedicated to 12 million Holocaust victims who suffered and died at the hands of Adolf Hitler and his Nazi regime', *www.nizkor.org/*.

communities such as 'café' communities, which are similar to both online/ offline communities and gaming communities. A 'café' community, much like an online/offline community, is centred on the ability to meet informally with other community members for discussion. Café communities may be distinguished from online/offline communities, though, by their independence from real-world events or activities. They are also distinct from gaming communities because there is no record of progress or offer of reward: café communities simply use the network to create gathering places for people to meet to discuss anything from politics to sex to the weather. Two of the most interesting online communities are 'knowledge' communities and 'creative' communities. Both are focused on the distribution of knowledge, but each in a slightly different fashion. Knowledge communities provide for the pooling of discrete information, by gathering information from a large network of individual members, to be fed into an extensive and informative database, which may then be used for the benefit of the community as a whole. They come in a variety of forms and vary from consumer knowledge communities such as Tripadvisor, which will tell you which hotels to book and which to avoid, and Review Centre, which gives customer reviews on a wide variety of consumer goods and services to professional communities such as mdchoice-.com, which offers both a general advice site to the public as well as a specialised support structure to medical professionals. Knowledge communities are one of only two online communities that use the power of the network to create, the other being the final community classification, which is the 'creative' community. 'Creative' communities differ from knowledge communities as they use the collective skills of the members of the community to create or develop a new product or service. Creative communities grew out of the collaborative nature of network development in its early years. In the 1970s, solutions to problems such as internetworking discrete networks such as ARPAnet, ALOHAnet and SATnet had been solved by a group of creative computer scientists working collaboratively.[82] Later projects like Richard Stallman's open source software venture – GNU – had harnessed the power of collaborative development.[83] Now creative communities are prevalent in the Open Source and Creative Commons movements. One of the best known of these projects is the Wikipedia project, which invites individuals to contribute to the creation of an extensive online encyclopaedia. The Wikipedia project was created in January 2001, and by the time of writing (April 2006) had over 2.6 million entries in 10 languages, including over 1 million entries in English. When compared with the 238-year-old Encyclopaedia Britannica, which has only around 120,000 entries[84] we see that Wikipedia has already achieved its aims of being 'the largest encyclopaedia in history, in terms of

82 See above, pp 64–70.   83 *www.gnu.org/*.
84 Information taken from Wikipedia, *http://en.wikipedia.org/wiki/Encyclopedia_Brittanica*.

both breadth and depth.'[85] This explosive growth has been achieved through the use of 'wiki', a web application that allows users to add and edit content directly, and through the contributions of thousands of members of the wiki creative community.

There are several additions or qualifications that should be added to this topology. The primary qualification is that rarely will a community fit into a single classification. Most communities, both online and offline, are complex and multifaceted and cannot be classified simply by reference to a single attribute of that community. Communities normally support several distinctive types of community interaction and will usually do so simultaneously. eBay, for example, is primarily a commercial community: members of the eBay community gather to engage in buying and selling of goods and the community regulates this activity through a thoroughly effective monitoring and feedback system. Errant members of the eBay community who disrupt the wider community in the pursuit of its primary purpose – either with intent through activities such as false listings, selling of counterfeit goods or failure to pay, or simply through poor practices such as unnecessary delays or making unreasonable demands – will have their actions recorded by, and publicised within, the community, through the addition of negative feedback to their publicly accessible member profile. This policy has been so successful that eBay members take exceptional pride in achieving and holding a positive feedback score of 100 per cent and those whose feedback drops below 90 per cent will find it difficult to do business within the community. But despite the obvious primacy of trading within the eBay community it is more than just a commercial community. eBay members also make use of an extensive and widely visited community forum where they can receive the advice of other community members about all aspects of eBay and about products or services,[86] or where you can simply get together in discussion.[87]

Thus when examining a community dynamic it is important to be able to distinguish the community's primary purpose from any secondary or derivative support systems within that community. Second, the primary purpose of a community may change over time. A community that starts out life as a gaming community may over time lose the gaming element to its interaction and migrate to become a café community; similarly, a café community or an online/offline community may migrate to become a commercial community. Thus communities are dynamic; in fact, a community that fails to adapt, risks stagnation. Third, communities may be affected in their pursuit of their purpose by the available technology. Thus although eBay may be a commercial community, many aspects of any commercial transaction on the eBay system

85 *http://en.wikipedia.org/wiki/Wikipedia.*
86 In this respect the eBay community is acting as secondary Knowledge Community.
87 In this respect the eBay community is acting as secondary Café Community.

are highly unsatisfactory: there is no opportunity to inspect the goods fully –
customers must 'make do' with pictures or videos; further, customers have no
guarantee that the goods they receive will be the same as those pictured, as
the seller may switch the goods before shipping. The major problem with any
eBay transaction though is a lack of synchronicity. When goods are purchased in the real world the buyer and the seller synchronise their transactions so that payment and delivery occur simultaneously. As an online trading
community this is impossible in eBay transactions: the buyer must pay before
delivery or the seller must risk non-payment. Extensive systems have been
developed by eBay to circumvent this problem. eBay suggests that all its
members use the PayPal payment system, which is owned by eBay and which
can be used to make a refund to the buyer should the seller fail to deliver or in
the event that the goods supplied fail to meet their description, and which
protects the seller from unauthorised chargebacks made by buyers. This system, although satisfactory for most transactions, is insufficiently secure for
high-value transactions and therefore eBay has developed an extensive menu
of payment options, including the use of escrow for high-value items. The
reason eBay has developed such systems is to alleviate some of the technological barriers that interfere with the ability of its members to pursue their
purpose. eBay is, of course, not the only community that suffers from this
problem; most communities suffer from bandwidth availability problems with
even extremely robust community hosts like ezboard and the BBC requiring
that members do not disseminate off-topic messages on boards or resubmit a
contribution to more than one discussion. It is important to remember therefore that when dealing with online communities the driving force behind any
decision of that community may not be its primary, or even secondary
purpose: it may be a simple reflection of the limits of the available technology.

With those qualifications in mind, how do online micro-communities regulate members? As we know, communities regulate by adopting a set of community values. In the informal, these values may be termed as standards: the
level of behaviour that is expected from the members of the community. In
the formal, these values may be termed as norms: the level of behaviour that
is required of members of the community and which, if not met, will lead to
the expectation of a rebuke or stronger response up to and including exclusion from the community. The norms or standards are normally enforced by
our everyday interaction with other members of the community who detect
deviation from that standard and who will apply the relevant sanction to
correct this deviation. Such detection and enforcement may itself take on one
of two forms: decentralised detection and enforcement, such as asking people
not to put their feet on seats in trains or rebuking someone who sneezes in
your direction without covering their face, and hierarchically enforced sanctions such as those of the Law Society or a local residents society where a
set of established norms is enforced by a management or disciplinary committee. We may call the first type of community regulation, pure community

regulation, while the latter is a community/hierarchy hybrid.[88] Both decentred and the managed communities may also be found in cyberspace. Most commercial communities are managed by the service provider: eBay for example may control listings and delist items in breach of the eBay prohibited and restricted items policy; similarly most online/offline communities and café communities employ moderators to ensure that listings, discussions and messages do not contravene policies of decency, privacy or rules against commercial listings. Conversely, most creative communities such as open source, creative commons and wiki are decentralised. There is no overall editor of the Wikipedia project. Although the originator of Wikipedia, Jimmy Wales, is described affectionately by contributors as their 'benevolent dictator',[89] he has no control over content or listings. If a member of the wiki community disagrees with an existing entry they simply edit it: there is no need to seek permission or to ask for third-party intervention. This open source editorial policy has been variously described as a social experiment in anarchy or democracy, and as an experiment in a Darwinian evolutionary process[90] and is not without problems.[91]

Another, and quite distinctive, form of decentred community are peer-to-peer (P2P) communities. These rather unique online communities are quite unlike the real-world communities we are familiar with. Their purpose is to facilitate a market in digital products and their design is streamlined for the simple exchange of digital information. As such they are commercial communities – their function to facilitate that most basic form of commerce: barter. This at least is how they may be described by supporters of P2P technologies and P2P communities. In truth they represent that least desirable of commercial community, the community of black market traders who seek to deal in illegal, or at least morally questionable, goods – no questions asked. Unlike their real-world counterparts, though, they are made up mostly of people who would describe themselves as law-abiding and morally upstanding. Trading digital information is seen as a victimless crime, and members of P2P communities would react in horror at the thought of being classified alongside black marketeers. P2P communities are deliberately structured in such a way as to protect their members from the intrusions of the

---

88 Chapter 2 and Murray, A and Scott, C, 'Controlling the New Media: Hybrid Responses to New Forms of Power', 2002, 65 MLR 491.

89 *http://meta.wikimedia.org/wiki/Power_structure*.

90 *http://en.wikipedia.org/wiki/Wikipedia*. See also Waldman, S, 'Who knows?' *The Guardian*, 26 October 2004.

91 'Edit wars': disputes between two or more contributors which lead to constant re-listing and re-editing of page content sometimes break out. Although troublesome in the short term these are usually only a minor problem in the medium/long term and in the event they continue they can usually be defused through mediation or arbitration, *http://en.wikipedia.org/wiki/Wikipedia:Edit_war*.

authorities. After the Napster decision[92] the designers of P2P software protocols took steps to remove any centralised access points from software kernels
such as Gnutella, meaning that, in software terms, there is little difference
between P2P systems such as Gnutella and more traditional networks such as
the web. Any social control exercised in P2P networks must come from within
the community itself.

Having drawn a distinction between these two types of online community
and identified the methodologies of control exercised within each, I hope,
now, to demonstrate the effectiveness, or otherwise, of these controls through
two case studies. The first involves community regulation within the centrally
managed eBay community while the other looks at the effectiveness of
community controls within a decentred P2P community.

Case study I: The Live 8 tickets scandal

An excellent example of the power of community regulation in the online
environment was seen in the summer of 2005. On 1 June 2005 the musician
and political activist Bob Geldof launched a series of simultaneous free concerts to be held on 2 July that year in five major cities.[93] The concerts were to
be called Live 8, and were designed to act as a focal point for a wide political
constituency to call for justice for Africa and the world's poor. They were
designed, alongside events to take place in Edinburgh on 6 July, including a
march of up to 1 million participants and a further free concert, to focus the
world's attention on decisions with regard to third-world debt and aid,
which were to be made by the G8 group of leading industrial nations at its
Gleneagles conference on 6 July.[94] At his press conference on 1 June, Geldof
revealed the mechanics of these concerts including the complex ticketing
arrangements. As the concerts were to be free, tickets for the main event in
London's Hyde Park would be allocated following a lottery. Entries to the
lottery would be made via text message with the details announced in the
national media at 8.00 am on Monday 6 June. Entry required answering a
simple question and texting your response to a number managed by the
mobile telecoms company O_2. The lottery closed at midnight on Sunday
12 June, by which time over 2 million entries had been received for the 72,500
pairs of tickets available. Notification of winning entries began on 13 June
and almost immediately some winners decided to cash in on their prize. By

---

92 *A&M Records Inc v Napster Inc* 114 F Supp 2d 896 (ND Cal 2000).
93 The five were Paris, Berlin, Rome, Philadelphia and London.
94 The G8 conference was to examine a plan from the Africa Commission to write off the debts
    owed to the world's richest countries by the world's poorest countries, to double the amount
    of high-quality aid offered to those countries by adding an extra $50 billion to available aid,
    and to make amendments to trade laws to allow those countries to build a future for
    themselves.

early morning on Tuesday 14 June over 100 pairs of tickets were on offer for sale on the eBay UK auction site, with early indications suggesting sellers could achieve a price of up to £1,000 for a pair of tickets. Bob Geldof responded to this news in his usual bullish way. He described those selling Live 8 tickets as 'miserable wretches who are capitalising on people's misery',[95] while eBay was described as 'acting as an electronic pimp' by allowing transactions.[96] Originally, eBay's management took an equally bullish stance. At 7.19 am on 13 June a posting from the 'eBay.co.uk Team' was made to the eBay.co.uk general announcements discussion board. It stated that eBay wanted to let its customers know that it would be permitting the sale of Live 8 concert tickets on eBay.co.uk. The message went on to explain that they were allowing the sale of these tickets because they believed that people could make up their own minds about what they buy and sell and that the reselling of charity concert tickets is not illegal under English law. The message went on to explain that eBay believed it to be a fundamental right for someone to be able to sell something that is theirs, whether they paid for it or won it in a competition. Simultaneously, eBay released a statement to the media reiterating this frame of mind and stating that 'a ticket to the Live 8 concert is no different from a prize in a raffle run by another charity and what the winner chooses to do with it is up to them'.[97] Throughout the remainder of 14 June an open dispute between Geldof, eBay members and eBay could be followed by anyone accessing the eBay.co.uk website. While Geldof threatened legal action and generally blustered to the press about the moral corruption of all those involved, a far more concerted campaign began. A group of eBay members formed themselves into a small group, calling themselves 'ticket-toutscum'. The group was organised through listings on the eBay.co.uk site and quickly set out to wreck the sale of Live 8 tickets by making hoax bids on all Live 8 ticket sales, such as the seller from Penzance in Cornwall who had 22 steady and rising bids on his/her pair of tickets rising from a five-pound starting price up to 900 pounds for their two tickets, who then found the next bid registered was for 10 million pounds (the maximum allowed by the eBay UK system) – their sale now being irrevocably disrupted. Although undoubtedly some sales were concluded, mostly using eBay's buy-it-now option, or by sellers requiring buyers to pre-list before bidding, the vast majority of sales were blocked by the action of this user group. While tickettoutscum was taking direct action against listings, the wider eBay community was making itself heard. Both through direct emails to eBay.co.uk and through postings on eBay message fora such as 'The Nag's Head', 'The Round Table' and 'The

95 Geldof calls for eBay boycott, *Daily Mail*, 14 June 2005.
96 eBay bans sale of Live8 tickets after 'electronic pimp' accusation, *The Independent*, 15 June 2005.
97 Geldof calls for eBay boycott, see above, fn 95.

Community Question and Answer Board', many among the community expressed a strong dissatisfaction with eBay's approach to the problem.[98] Throughout the day the discussion fora buzzed with debate between the majority of eBay members who saw the sale as immoral and damaging to eBay's reputation and a vocal and vociferous minority who argued that the tickets should be treated as similar to any other goods, and who drew comparisons with charity wristbands, which are freely on sale on the eBay.co.uk site.

By 7.34 pm on 14 June a winner was declared. Doug McCallum, Managing Director of eBay (UK) Ltd. posted the following message on the general announcements discussion board.

> Dear all
>
> Today you have made it very clear to us that our previous decision to allow the sale of LIVE 8 tickets on eBay.co.uk was not one that the vast majority of you agreed with. As a result of this clear signal from the Community we have decided to prohibit the resale of LIVE 8 tickets on the site.
>
> Although the resale of tickets is not illegal, we think that this is absolutely the right thing to do. We have listened to the views you expressed on the discussion boards and in the many emails you have sent to us. We shall be working over the next few hours to remove all LIVE 8 ticket listings from the site.
>
> Thanks for taking the time to contact us and make your views heard,
>
> Regards
> Doug McCallum
> Managing Director, eBay (UK) Ltd.
> On behalf of the whole eBay.co.uk team

As with their posting that morning eBay made a simultaneous announcement to the press, stating that 'We have listened to eBay's community of users and the message has been clear. They do not want the tickets to be resold on the site. Once we are made aware of any Live 8 tickets being resold, they will be taken down.'[99] The community had spoken, and the eBay team had been

---

98 A sample message noted 'Dear eBay, You have a couple of days at most to stop allowing these ticket auctions or you lose me as a customer. My last act will be to send abusive messages to the people auctioning the tickets. I will do this until you cancel my ID. It is NOT GOOD ENOUGH to hide behind UK law saying that it is not against the law to act as an agent in this. For goodness sake, have some corporate self-respect.'
99 eBay bans sale of Live8 tickets, see above, fn 96.

forced to take action. What is perhaps most important about this tale is that it was not the actions of Bob Geldof, or the response of the media which forced this change, although no doubt eBay.co.uk was concerned about its public image. Neither was it the direct action of the tickettoutscum group which caused this change in policy; it was the concerns and complaints from the wider eBay community. Throughout, eBay handled the issue with perfect balance, and the ultimate decision it appears rested with the community. Threats of legal action and media intimidation were strongly resisted at a time when eBay.co.uk believed its community wished to support the sale of these items. The way the eBay management team handled the affair is most instructive. eBay it should be remembered, is a commercial community and therefore it should be assumed in the first instance that its community would seek to support the free market in all goods and services. When a minority of members acting under the name, tickettoutscum, began to take direct action the eBay management team, quite rightly, refused to be cowed by this. The accounts of those who took such action were suspended under eBay's 'acceptable use policy' with members being told that they had made a 'deliberate attempt to disrupt an auction' and as such were in breach of the policy.[100] What eBay did throughout, though, was listen to its community. When it became clear that the majority of the eBay.co.uk community did not support the original stance of the eBay.co.uk team, action was taken to remedy the situation. The UK eBay community may be one driven mostly by commercial imperatives. It may be a fractured community of individuals who mostly are involved in individual transactions, but what was demonstrated on 14 June 2005 was that any virtual community can draw together when there is a strong common interest in favour of a particular outcome. But to agree what is desirable and to achieve it are two different things as will be demonstrated below.

Case study II: Peer-to-peer and porn

Peer-to-peer (P2P) communities are loosely knit communities; membership of the community is usually obtained by simply installing a small piece of software on your computer – you need give no information about yourself except for the location of your computer on the network.[101] While most people's experience of P2P communities is to trade music, movies or software, there is a darker side to P2P life. All major P2P protocols may be used to trade any type of digital file and one area of growing concern is the proportion of pornographic material available on P2P networks, and in particular the availability of child pornography. Dealing with pornography is

---

100 Live8 eBay high bidders banned, *BBC News*, 15 June 2005, available at: *http://news.bbc.co.uk/1/hi/entertainment/music/4095464.stm*.
101 In other words your IP address.

mostly a matter of controlling supply. Pornography is, on the whole, not illegal.[102] Rather, it is a controlled product much like alcohol, tobacco, knives or solvents. The main reason we control the supply of pornography is to prevent children accessing pornography while they are at an immature stage of development. A particular danger of all networks, including P2P networks, is that without the badges of physicality, which we all carry in the real world, children may gain access to this content. Such concerns led to the Communications Decency Act[103] and the Child Online Protection Act[104] being promulgated in the US and led to Congressman Joe Pitts of Pennsylvania introducing the Protecting Children from Peer-to-Peer Pornography Bill in Congress in July 2003. This Bill, if passed, would: have required P2P distributors to give notice of the threats posed by P2P software; distribute P2P software to a minor only with a parent's consent; comply with the Children's Online Privacy Protection Act when gathering information from children under age 13; ensure that the software could be readily uninstalled; and ensure that the user's computer could not be used as a super node, or that software could not disable or circumvent security or protective software, without consent. The Bill failed to get past its first reading, but the fact that it was introduced demonstrates a fear in the wider population that pornography is widely available in P2P communities and that children are routinely exposed to such content.

This impression has been tested by several academic studies and these seem to bear out the fears of the wider community. In their 2004 paper *Measurement Study of Shared Content and User Request Structure in Peer-to-Peer Gnutella Network*,[105] Przemyslaw Makosiej, German Sakaryan and Herwig Unger of the University of Rostock found that the top 30 keywords used in Gnutella searches were dominated by pornographic terminology with the three most common keywords being 'XXX', 'Sex' and 'Porn'. Perhaps more worryingly were the appearance of 'Girl', 'Young' and 'Lolita' among the top 30.[106] The authors postulated that the high proportion of what they termed 'x-rated keywords' was 'because they represent accumulated interests

---

102   Under s 1 of the Obscene Publications Act 1959, 'an article shall be deemed to be obscene if its effect or (where the article comprises two or more distinct items) the effect of any one of its items is, if taken as a whole, such as to tend to deprave and corrupt persons who are likely, having regard to all relevant circumstances, to read, see or hear the matter contained or embodied in it'.

103   The Communications Decency Act was Title V of the Telecommunications Act of 1996, Pub. LA. No. 104–104, 110 Stat. 56 (1996).

104   H.R.3783 (1998).

105   Makosiej, P, Sakaryan, G and Unger, H, 'Measurement Study of Shared Content and User Request Structure in Peer-to-Peer Gnutella Network', *Proceeding of Design, Analysis, and Simulation of Distributed Systems* (DASD 2004), Arlington, USA, April 2004, 115, available at: *www.teo.informatik.uni-rostock.de/~gs137/articles/sakaryan04measurement.pdf*.

106   Ibid, table 5.

of the users in x-rated materials, which is much higher than interests to any particular file (music or video)'.[107] In other words the results are biased as users searching for pornography tend to use generic terms like 'Porn' or 'XXX', whereas users searching for music or films tend to use specific search terms such as 'Kaiser Chiefs' or 'Franz Ferdinand', not general terms like 'Guitar' or 'Rock'. While this is no doubt true, another study by KY Chan and SH Kwok of The Hong Kong University of Science and Technology has found that even accounting for this bias, searching for and downloading pornography makes up a substantial proportion of P2P activities. In their paper *Information Seeking Behaviour in Peer-to-Peer Networks*,[108] they found that sexual search queries made up 23 per cent of Gnutella searches, making it the second most popular search query after entertainment files, which made up 53 per cent of queries.[109] The widespread availability and popularity of pornographic content within P2P communities cannot therefore be denied.

The availability and popularity of such content is arguably not even of relevance at this point of this book as its high popularity and availability suggests that it is not seen to be particularly improper within P2P communities. In fact, the fact that pornographic content remains widely available within P2P communities may be seen to be a measure of the communities' resolve to protect their values and interests, despite attempts at outside interference from lawmakers and the media. The purpose of these communities is, after all, trade, not the protection of minors. What is much more interesting is not the response of these communities to the availability and trade in pornography – it is how they respond to child pornography. Child pornography differs considerably from mainstream pornography. Possession of child pornography is a criminal offence in all industrialised nations,[110] and the criminalisation of child pornography in relation to online distribution has recently been reiterated in Article 9 of the Convention on Cybercrime.[111] There is no doubt that the vast majority of the P2P community must find the trade in child pornography to be immoral and unacceptable, even if they do not feel the same way about mainstream pornography.[112] Despite this, a

---

107   Makosiej, Sakaryan and Unger, above, fn 105, p 122.
108   Chan, K and Kwok, S, 'Information Seeking Behaviour in Peer-to-Peer Networks: An Explanatory Study', *Proceeding of the International Symposium on Digital Libraries and Knowledge Communities in Networked Information Society* (DLKC '04), Tsukuba, Ibaraki, Japan, March 2004, available at: *www.kc.tsukuba.ac.jp/dlkc/e-proceedings/papers/dlkc04pp40.pdf*.
109   Ibid, p 4.
110   In the UK under s 1 of the Protection of Children Act 1978 (as amended).
111   Council of Europe, ETS No. 185, Convention on Cybercrime, Budapest, 23 November 2001.
112   This proposition is supported by the terms of the P2P United, Software User Advisory, which states that 'The member companies of P2P United believe strongly that we all must stamp out child pornography' and asks users to report incidences of Child Pornography to 'the relevant authorities'.

search carried out by the author in July 2005 using a popular Gnutella client revealed that the term 'Lolita' returned 769 files – mostly video files – while the term 'underage' returned 334 files – again mostly video files – and searches with two search terms frequently used by individuals who trade in child pornography returned 465 and 298 files, again with most returns being video files. The question is, why, two years after the US General Accounting Office highlighted the problem, by revealing that 'by using 12 keywords known to be associated with child pornography we identified 1,286 items and further determined that 543 (about 42 per cent) of these were associated with child pornography images',[113] could a simple search using only four keywords provide a much higher return of 1,866 files?[114] It may simply be that the community takes no interest in the actions of those who trade such content, or feel there is little they can do, yet this does not seem to be the case.

One discussion group on the Zeropaid Bulletin Board discusses at length how to deal with child porn with the discussion perhaps illustrating why the community is failing to take effective action.[115] The discussion is opened by one member 'Cyrill666' asking 'how can we, as a community, remove this unnecessary and vile type of file from the file sharing environment? I know we can ignore files using keywords etc, but that's turning a blind eye in my opinion, not actually dealing with the problem.' The discussion then rages among community members, some suggesting that the community should simply turn a blind eye as exemplified by one member 'and1_corey' who replies, 'child porn is just sick . . . it's only going to get worse because the morality of mankind is only getting worse . . . the best we can do is try to ignore it and go on as normal civilized people', while others plead for a freedom from external intervention, such as 'jabba | xtra', who echoes the statement of 'and1_corey', but with a slightly different sentiment: 'yep, just ignore it, leave your filters alone, and all will be fine. It's not right, but what are we gonna do about it. I sure as hell dont want "p2p police". Just ignore it and go on like before.' Some community members though suggest positive action. One proposal is for a digital 'wall of shame', which would allow users to post the IP address of those who trade in child porn, but one user 'cracker-jacker' feels this is not a practical solution, noting 'the wall of shame existed before on zeropaid. They probably do need to be putting it back again but only by user names and the reason putting the names might not work, is those individuals who are sharing the child porn might change their names too. So, to add names wouldn't be a good idea, either. They wouldn't be able

113   United States General Accounting Office, *File-Sharing Programs: Peer-to-Peer Networks* Provide Ready Access to Child Pornography, GAO–03–351, February 2003, p 2.
114   Although many of these returns were duplicates there were over 900 unique file extensions. While the content of these files remained uninvestigated by the author, there is no doubt a high proportion of them would have contained actual images of child pornography.
115   The discussion may be viewed at: *www.zeropaid.com/bbs/archive/index.php/t–2619.html*.

to post IP addresses, because IP addresses are frequently interchangeable. For example, dial-up users, connect to the net with a new IP address each time they dial in. So what if some individual who had nothing to do with the sharing of the child porn has to have his IP addresses targeted even if it wasn't them sharing child porn, this would be unfair to target innocent people. DSL users, also have their IP addresses changed, also, I know this, because my IP addresses change all the time, when I reset the modem. Now, posting names of individuals, on the wall of shame, might help, but, think about this, they will always be able to change there (sic) name.' Another solution is posed by another user 'gorphon' who suggests 'one thing that would perhaps work – over a long period of time – would be a grassroots style campaign among the file sharing community, finding the IP addresses of known distributors and downloaders of child porn (pretty easy with netstat), tracking it down to the host (online trace and look up tools) and bombard the isp (usually abuse or admin@whateverisp.com) with complaints of transfer/ hosting of pornographic images of children. I suspect most hosts would drop these people just as quick as they could, especially if they recieved a large number of complaints.' This idea initially seems to be well received among the discussants, but one member 'havoc' replies 'we can't just pretend this isn't happening we have to take an action what action I don't know. Even if they ban users they'll always come back as another isp.' More ideas continue to flow among the group. An idea for a P2P Police is proposed by 'WhitePony' while the idea of a 'vote to ban' content through reporting buttons is suggested by 'Jared592', while most radically 'ashep612' advocates abolition of P2P networks saying 'I would gladly give up filesharing to stop child porn'. In the end though all discussions come to nothing with two postings dealing with virtually all the proposals. The first is a lengthy posting by 'klimt da man':

One: the idea of P2P police? Even if it were desirable, forget about it. Pipe dream. How many files? We're talking software-based or nothing.

Two: it's always been around this is nothing new? I think that's like saying pirating music has always been around so P2P doesn't change it ... but it does, people. I agree – I'd be willing to give up P2P to stop child porn – but it won't. However, thinking the situation is unchanged by P2P is very naive, I would say. Now they have unprecedented potential to share this stuff.

Three: Had my first look at the wall of shame – not a thorough look but a glance – and I can't see how posting IP addresses will do any good. I mean, would I recognize my own IP address? And as people keep pointing out, user names can be changed in a blink. I like the idea, but for two problems: i) ambiguity – we can't have people getting nailed for some mistake – reminds me of the whole situation in child care and counseling, how teachers can't counsel with an office door closed cause they're damn scared of a false accusation. This stuff has heavy repercussions and people tend to go into witch-hunt mode (understandably, but still dangerous); ii) I like the idea of something unambiguous like specific files agreed upon with a warning to the user and a post to the wall – but then you have entrapment – luring people who might not commit the crime to do so. Same thing if I go around offering crack. Different if they come looking, but if I offer, I'm preying on weak wills, which is why the law (in my country, at least) covers that.

The second is a shorter posting from 'HansG' dealing with the 'vote to ban' proposal:

The Vote To Ban idea is a bad one. Very naive. It all comes down to this: restrict those who share child pornography, and it will come back in bigger quantities under the wrong names.

We don't need any police, the Internet is a mess and that's the nature of the beast. Give anyone control of content and you defeat the purpose of it. Imagine the Vote To Ban idea hacked. Say I can Vote To Ban you, but my vote counts for 10,000 votes . . . you are the weakest link, goodbye. Even worse: the RIAA would make it top priority to have this thing hacked, a great way to boot you off . . . forever. Now just automate and it will boot off any user. Do you see where I am going? Your ideas – god forbid a programmer to implement them – could well result in the destruction of the thing you are trying to protect.

This discussion strand encapsulates perfectly why there has been a failure on the part of the P2P communities to take concerted action against those who use their communities' networks to trade in child pornography. Although there is a clear sense of distaste, anger and even outrage at the actions of the minority, there is no consensus on how to act. This is because in highly decentred online communities such as P2P communites there is little sense of ownership. In the 86 postings in this thread the word community only appears four times. Twice it is used in the abstract to refer to all users of P2P software and twice in the true sense as a group of people with a common aim or purpose. The lack of use of the word suggests there is no real community feeling among the posters and as a result there is no one willing to take ownership or action on behalf of the community. Some ideas, such as the report to ISPs and the vote to evict ideas may have been fruitful, given some further development, but instead, once the problems are pointed out they are abandoned as no one is willing to take the idea and develop it beyond a talking point. This reflects the weakness of online communities: the lack of physical proximity removes much of the signals between individuals that form the bonds of community identity. This is particularly damaging in decentralised communities where there is no one in a position of authority to take control of a project. Without someone to take ownership of its ideas it is extremely difficult for an online community to reach consensus and even where some form of consensus is achieved, it is equally difficult for the community to take positive action.

Where next for cybercommunities?

These two case studies prove to be most illuminating. The clear success of the community in achieving a change to the regulatory settlement in the Live 8 case is equally mirrored by a similar lack of success in the P2P example, but why? The answer is in the nature of the communities. The eBay community is by nature a Lockean community. It is one where the community has by covenant agreed to relinquish power to the eBay management. It may be claimed that the analogy with a Lockean covenant is somewhat false, as those who seek to use eBay have little choice but to accept the terms offered to them by the eBay management team, but to do so would obfuscate the underlying reliance that the eBay management have upon community support. In much the same way that democratic government needs at least the tacit support of its people, the same may be said of eBay, should eBay introduce terms and conditions felt to be unreasonable, and should they refuse to change them, the members of the eBay community would migrate elsewhere. It is this which ultimately forced the eBay UK management team to delist auctions offering for sale the Live 8 tickets. The internet is filled with Lockean communities, from commercial communities such as eBay to online/offline communities and café communities, which make use of list owners and moderators, to gaming communities that are controlled by designers and 'Wizards'. In all these communities the strength offered by the centralised structure of the community allows for effective community regulation. Democratic governance is effective in cyberspace with the network of members of the community acting as the community's detectors, who then report any errant behaviour to the moderators or managers of the community, who act as effectors, taking such corrective action as is necessary. By comparison, P2P communities are imperfect Rousseauen communities. There is no divorce of power from the community in decentralised online communities. According to Rousseau's normative social contract theory, decisions should be made from within the body of the community for the collective benefit of all members of the community, but it appears this is not occurring. Why? The answer would appear to be that such communities are simply too large and too diverse to be subject to Rousseau's contract. As we saw earlier, Rousseau's strong form of democracy is also only possible in relatively small communities. The people must be able to identify with one another, and at least know who each other are. They cannot live in a large area, too spread out to come together regularly, and they cannot live in such different geographic circumstances as to be unable to be united under common laws. Decentred online networks such as P2P networks are simply too remote and too anonymous to come together to form a common will. Members of these communities simply do not feel a community spirit, and do not fear community sanctions. With no one to give moral guidance, the community simply fragments into smaller and smaller micro-communities, such as those who wish to trade music, those who wish to trade films and those who wish to trade child pornography. Each of these narrow community interests simply ignore the others so long as they do not impact directly on their activities. Thus we find comments such as 'just ignore it and go on like before' become commonplace. Thus although community regulation is possible in cyberspace, it appears it is only effective as a community/hierarchy, or Lockean, hybrid.

Chapter 6

Competition and indirect controls

The Internet isn't free. It just has an economy that makes no sense to capitalism.

Brad Shapcott

In 1983 Frank Easterbrook had probably not considered the network of computers that was growing around him while he worked at the University of Chicago. While it would be another 13 years until his attack on the network, and those who choose to study legal-regulatory constructs within it,[1] Easterbrook's work that year shadowed some of the foundations of his later assault and provides a useful background to this chapter.

Central to all of Professor Easterbrook's work is the claim, common to all those who preach normative law and economics theory, that efficiency is the desired outcome of legal rules and legal settlements and that the market is the most desirable route to such efficiency.[2] He chose that summer to elucidate this claim with respect to two diverse subject areas: the first being the use of prosecutorial discretion, plea bargaining and sentencing discretion in the criminal process,[3] the other being competition law and the role of states in the two-level regulatory structure.[4] In the former paper, Professor Easterbrook patiently explained the market value and thereby efficiency of allowing for flexibility in the process of criminal justice. He noted that 'prosecutorial

---

1 See above p 9, and Easterbrook, FH, 'Cyberspace and the Law of the Horse', 1996, U Chi Legal F 207.
2 See eg Stigler, GJ, 'The Theory of Economic Regulation', 1971, 2 *Bell Journal of Economics* 3; Peltzman, S, 'Toward a More General Theory of Regulation', 1976, 19 *Journal of Law & Economics* 211; Posner, R, 'Theories of Economic Regulation', 1974, 5 *Bell Journal of Economics* 335. For a general overview see Posner, R, *Economic Analysis of Law*, 6th edn, 2002, Frederick, MD: Aspen Publishing.
3 Easterbrook, FH, 'Criminal Procedure as a Market System', 1983, 12 *Journal of Legal Studies* 289.
4 Easterbrook, FH, 'Antitrust and the Economics of Federalism', 1983, 26 *Journal of Law & Economics* 23.

discretion, plea bargaining and sentencing discretion may be understood as elements of a well-functioning market system. [They] set the price of crime, and they set it in the traditional market fashion'.[5] Easterbrook proposed that such a resolution is most efficient[6] when compared to externally imposed regulatory settlements[7] and as a result suggested that 'the interaction of judges, prosecutors and defendants in a market price-setting system tends to obtain the objective of obtaining the maximum deterrent punch out of whatever resources are committed to crime control'.[8] In the latter paper Professor Easterbrook was even more forthright in his views. Regulation, we were told, displaced competition. 'Displacement is the purpose, indeed the definition of regulation. Limitations on the number of taxicabs, licensing of barbers and dentists, health and safety codes, zoning laws and price supports for milk depend on the belief that competitive markets should be replaced with something else.'[9] This he argues is inefficient, for while in some cases 'legislation may be justified as necessary to correct "imperfections" in markets . . . in most cases legislation is designed to defeat the market altogether'.[10] This, it is further argued, is undesirable for as 'evidence shows regulatory laws owe more to interest group politics than to legislators' concern for the welfare of society at large'.[11] Thus in returning to the theme of normative law and economics theory the most efficient regulatory settlement was, in Easterbrook's view, one where there is no regulatory intervention by states.

The value of discussing these papers, written some 23 years ago and dealing with diverse and complex regulatory settlements that are quite unrelated to the topic of this book may, rightly, be questioned. Their importance though is in the core values expressed by Professor Easterbrook, and by all others who count themselves as members of the Chicago or Normative School of Law and Economics. It is the belief of this school of thought, as demonstrated by the quotes used above, that regulatory intervention should be pursued only as a last resort in cases of 'market imperfections' and that any regulatory intervention should follow the rule of Pareto efficiency.[12] Thus it may be argued that the most efficient regulatory settlement for cyberspace

5 Easterbrook, see above, fn 3, p 289.
6 'If transaction costs are not too high, the system yields efficient results in the sense that further moves cannot improve anyone's lot without making someone else worse off.' Ibid.
7 'In a regulatory system people may trade only on terms, or at times, laid down by third parties. Trading under regulation leads to an efficient result only if the regulator has set the same price that parties would have reached in their own bargaining.' Ibid.
8 Ibid, p 290.   9 Easterbrook, see above, fn 4, p 23.   10 Ibid.   11 Ibid.
12 Pareto efficiency states that given a set of alternative allocations and a set of individuals, a movement from one alternative allocation to another that can make at least one individual better off, without making any other individual worse off is a Pareto improvement or Pareto optimisation. An allocation of resources is Pareto efficient or Pareto optimal when no further Pareto improvements can be made.

is one that is dictated by the market, or to phrase it slightly differently, competition-based regulation is most desirable. This chapter will examine this thesis.

Identifying the relevant (cyber)market

There can be little doubt that digitisation and networking have had a profound effect on the world economy and markets. Economists and business leaders from around the globe have rushed to catalogue the effect that the transition from 'atoms to bits'[13] is having on the markets for products, services and information. Some, such as Erik Brynjolfsson of the Center for eBusiness at MIT, have chosen to focus on how digitisation has freed markets from the environmental constraints of the real world. This approach has led to the theoretical 'frictionless economy': a place where transaction costs, information uncertainties and production inefficiencies are reduced as far as possible, leading to, as near as is possible, a state of perfect competition.[14] Some focus on the development of an economy built upon transactions in information.[15] These commentators, led by Danny Quah of the London School of Economics (LSE), write about the theoretical 'weightless economy', an economy built entirely on information and informational products such as ICTs,[16] intellectual property, electronic libraries, databases, and bio-informational resources, including carbon-based libraries.[17] Others, including business leaders such as Bill Gates, choose instead to examine the wider implications of digitisation on traditional market transactions, including how digital media and digital products and productivity may lead to a re-evaluation of classical economics.[18] Finally, there is a group of economists and sociologists who examine the costs rather than the opportunities of the digital economy. This group focuses on the growing 'digital divide' that

13 Negroponte, N, *Being Digital*, 1995, Chatham: Hodder & Stoughton.
14 Amable, B *et al. Internet: The Elusive Quest of a Frictionless Economy*, 2003, available at: *www.frdb.org/images/customer/copy_0_rdb_final_draft_2.pdf*; Brynjolfsson, E and Smith, M, *Frictionless Commerce? A Comparison of Internet and Conventional Retailers*, 1999, Mimeo: MIT Sloan School; Hagel III, J and Singer, M, 'Unbundling the Corporation' *Harvard Business Review*, March/April 1999 133.
15 This is not to be confused with the knowledge-based economy that simply proposes that knowledge matters for economic performance.
16 Information and Communications Technology.
17 Quah, D, *The Weightless Economy in Economic Development*, 1999, CEP Discussion Paper No. 417; Coyle, D, *The Weightless World: Strategies for Managing the Digital Economy*, 1997, London: Capstone; Quah, D, *The Invisible Hand and the Weightless Economy*, 1996, CEP Occasional Paper No.12.
18 Tapscott, D, *The Digital Economy*, 1995, New York: McGraw-Hill; Gates, B, *Business at the Speed of Thought: Succeeding in the Digital Economy*, 2000, Harmondsworth: Penguin; Abramson, B, *Digital Phoenix: Why the Information Economy Collapsed and How It Will Rise Again*, 2005, Cambridge, MA: MIT Press.

has opened up between technology-rich western states and technology-poor African and Asian states, and on the growing divide within states between the professional classes with stable and fast internet access, and the working class, in particular immigrant communities, where access may be unstable, slow and difficult to obtain.[19]

More recently a group of eminent cyberlawyers and computer scientists have come together to specifically examine the role the market plays in encouraging human creativity and the production of intellectual products within the digital environment. This group, which includes Lawrence Lessig, James Boyle,[20] Hal Abelson[21] and Eric Eldred,[22] have formed the Creative Commons (CC): 'a nonprofit organisation to promote the creative re-use of intellectual and artistic works, whether owned or in the public domain, by empowering authors and audiences.'[23] They are attempting to introduce a new set of values within the narrow market for copyright works in digital form. They believe that the debate over control of creative goods has become polarised between those who follow market imperatives, who argue that control is central to the exploitation of such works,[24] and those who argue for complete creative freedom, but who leave such works vulnerable to exploitation.[25] Members of the CC community believe that copyright owners have responded negatively to the opportunities and threats of digitisation by constructing greater barriers (legal and design), which prevent the exploitation of creative works by the community at large. As noted by Lawrence Lessig:

> This rough divide between the free and the controlled has now been erased. The Internet has set the stage for this erasure and, pushed by big media, the law has now affected it. For the first time in our tradition, the ordinary ways in which individuals create and share culture fall within

19 Norris, P, *Digital Divide: Civic Engagement, Information Poverty and the Internet Worldwide*, 2001, Cambridge: CUP; Warschauer, M, *Technology and Social Inclusion: Rethinking the Digital Divide*, 2004, Cambridge, MA: MIT Press.

20 James Boyle is William Neal Reynolds Professor of Law at Duke Law School and author of *Shamans, Software and Spleens: Law and the Construction of the Information Society*, 1997, Cambridge, MA: Harvard UP.

21 Hal Abelson is Class of 1922 Professor of Electrical Engineering and Computer Science at MIT and co-author of *Structure and Interpretation of Computer Programs*, 2nd edn, 1996, Cambridge, MA: MIT Press.

22 Eric Eldred is editor and publisher of Eldritch Press, a free online book website at *www.ibiblio.org/eldritch*. He is better known as lead plaintiff in *Eldred v Ashcroft* 537 US 186 (2003) an attempt to overturn the 1998 Sonny Bono Copyright Term Extension Act.

23 Taken from *http://creativecommons.org/*.

24 Eg Landes, W and Posner, R, *The Economic Structure of Intellectual Property Law*, 2003, Cambridge, MA: The Belknap Press.

25 Eg discussion in Geist, M, Piercing the Peer-to-Peer Myths: An Examination of the Canadian Experience, 2005, 10(4) *First Monday*, available at: *www.firstmonday.org/issues/issue10_4/geist/index.html*.

the reach of the regulation of the law, which has expanded to draw within its control a vast amount of culture and creativity that it never reached before. The technology that preserved the balance of our history – between uses of our culture that were free and – uses of our culture that were only upon permission – has been undone. The consequence is that we are less and less a free culture, more and more a permission culture.[26]

The CC project is seeking to reverse this process and to revive balance, compromise and moderation in copyright: to provide what it calls 'the driving forces of a copyright system that values innovation and protection equally'. To achieve this aim, CC has created a menu of licences, which authors of original content may use in place of traditional copyright settlements. These include the 'attribution' licence that allows others to copy, distribute, display and perform the author's work (and derivative works), but only if they give them credit; the 'non-commercial' licence, which lets others copy, distribute, display and perform the work, but for non-commercial purposes only; and the 'share alike' licence, which allows others to distribute derivative works only under a licence identical to the licence that governs the original work.[27] The idea behind these licences is to create public goods while acknowledging the creative effort of the author. The CC is an innovative and exciting project designed to capture the creativity of the internet community and to channel this creativity in the hope of generating further and better public goods: it is, as Lawrence Lessig has said, a community response to Apple's Rip, Mix & Burn advertising campaign.[28] It is though, a project that finds itself under assault from all sides in the highly competitive marketplace for digital creative goods: file-sharers and digital pirates do not recognise a licence, however structured, while traditional content distributors continue to argue that recognition, and protection, of the traditional copyright model is essential to allow the market in digital content to mature. As a result, the market for creative content is currently in a high degree of regulatory flux, and for that reason will be used as a case study in this chapter.

Protecting bottles[29]

Creative works have always occupied a peculiar legal and economic niche. Although we talk of intellectual *property* rights, creative rights such as

---

26 Lessig, L, *Free Culture: How Big Media Uses Technology and the Law to Lock Down Culture and Control Creativity*, 2004, New York: Penguin, p 8.

27 For a full explanation of these CC licences see *http://creativecommons.org/about/licenses/*.

28 Bridges, M, 'Rip, Mix, and Burn Lessig's Case for Building a Free Culture', 2004, *Berkman Briefings*, available at: *http://Cyber.law.harvard.edu/briefings/lessig*.

29 This section heading is taken from John Perry Barlow's classic article, 'The Economy of Ideas: Selling Wine Without Bottles on the Global Net', *Wired 2.03*, March 1994.

copyright sit uncomfortably with the traditional foundations of property theory.[30] Property usually assumes an exclusive right of use, something which may be seen to be derived from the physical nature of traditional property (that is a pen by its nature cannot be divided and thus can only be used by one person at a time). Conversely, creative works, as famously noted by Thomas Jefferson, have no such restriction for 'he who receives an idea from me, receives instruction himself without lessening mine; as he who lights his taper at mine, receives light without darkening me'.[31] The nature of creative works meant that when the concept of exclusive property in land and goods was developed, there was no parallel for works of the mind. Throughout the intellectual highs of the ancient Greek and Roman worlds and the intellectual dark ages that followed, there was no exclusive property in the spoken word. During this period creative works were distributed by travelling storytellers, often wandering minstrels, who would travel from town to town and engage in singing, acting, storytelling and comedy. With most citizens unable to read, the minstrel was often the only way for individuals to learn the stories from the Scriptures, or to hear news of recent events such as famous victories in battle. The concept of anyone 'owning' such knowledge would have been quite impossible for the average medieval European to imagine. At this time, the only information that was in the exclusive control of any individual or group was the Word of God, which, by canonical law was only to be preached by the Church, and only in the language of the Church, which was Latin. In part, the Church retained sole ownership of this information because only the Church authorities had the necessary resources to reproduce and distribute religious texts, which had to be copied by hand.[32]

As we have seen though, environmental regulation is often overcome by technological advances and in the 1440s one of the greatest technological advances made by mankind altered forever the delicate regulatory balance that had existed for hundreds of years between creative performers, the guardians of the religious (and secular) texts and consumers. The invention was, of course, Johann Gutenberg's movable type printing press, an invention that allowed Gutenberg to revolutionise the reproduction of creative works. The first product Gutenberg turned his invention to was, quite naturally, the mass production of the most in-demand text of the day: the Bible. It is believed that Gutenberg produced about 180 copies of his printed and illuminated Bible over a period of three years: a number that seemed impossible for a society which, from time immemorial, had been forced to produce

---

30 For discussions on property theory see Waldron, J, *The Right to Private Property*, 1990, Oxford: Clarendon Press or Murphy, WT, Roberts, S and Flessas, T, *Understanding Property Law*, 2nd edn, 2003, London: Sweet & Maxwell.

31 Letter to Isaac McPherson, 13 August 1813. Full text available at: *http://odur.let.rug.nl/~usa/ P/tj3/writings/brfljefl220.htm*.

32 It is estimated that a single handwritten Bible could take a single monk 20 years to transcribe.

copies of written works by hand. Gutenberg in fact produced his Bibles in about the same time it would have taken a single team of scribes to produce one copy of the same text in a Scriptorium. Thus his machine did roughly the work of 180 teams of scribes. With this success under his belt Gutenberg began the trade in secular printed texts, including several editions of *Ars Minor*, a portion of Aelius Donatus's textbook on Latin grammar, but despite his genius for invention, Gutenberg's career as a publisher was soon to fail. A loan from his partner, Johann Fust, was called in and Gutenberg, despite his early successes, could not pay. Fust took over Gutenberg's presses, leaving Gutenberg to live a spartan existence, subsidised by the Archbishop of Mainz until his death in 1468. Despite his personal failure though, Gutenberg's invention was about to change the face of Europe. His press spread quickly, and with publishers setting up in all the great cities of Europe,[33] the printed word began to travel across the continent. It fed the growing Renaissance with an increase in literacy and since it greatly facilitated scientific publishing, was a major factor in originating the scientific revolution. Arguably though the most wide-reaching effect of Gutenberg's invention was to be felt in England. In 1476, a Gutenberg-style printing press was set up in Westminster by William Caxton. From here, Caxton produced books not in Latin, the language of the Church, but in English, the language of the people. Caxton had in fact published the first book in English, *Recuyell of the Historyes of Troye*, in Bruges the previous year, but it was at his Westminster Press that some of the most important early English texts, including Sir Thomas Malory's *Le Morte Darthur*,[34] and Geoffrey Chaucer's *The Canterbury Tales*[35] were published. This was a major breakthrough: continental printers had focused their trade upon the Latin speaking canonical market, this being the best way they felt to recoup their extensive overhead costs. Caxton, by focusing on a narrower market, that small group of prosperous English speakers who were looking for books written to educate or entertain in their own language, had created the secular book publishing industry.

The effect of Caxton's innovation was to cause a move from the oral tradition, where creative works were unfixed, to the print era where the primary methodology for the distribution of creative works would come through fixed media such as books, newspapers, magazines and later musical and videographic recordings. This progression caused a high degree of regulatory flux, which was not dissimilar to that which we see today in relation to digital media, which is caused by the retroversive trend of decoupling creative works from their carrier medium – in a sense a return to the pre-1450

33 For example we know that a printing press was built in Venice in 1469 and that in 1476, a printing press was set up in Westminster by William Caxton.
34 Published by Caxton in 1485.
35 First published by Caxton in 1476.

settlement. The post-Caxton regulatory flux first became apparent when, much like today, unauthorised copies of the most popular works began to be produced and circulated by unlicensed printers seeking to free-ride upon the popularity of these works. To deal with this problem the Guild of Writers of Text-letters, Lymners, Bookbinders and Booksellers, which had been formed in 1403 in the City of London, formed a publishing cartel.[36] The Guild passed an Ordinance that made it an offence to print a book before showing it to the Wardens of the Mayor of London for approval, and then registering such approval in the Register of the Guild.[37] The Wardens were in turn charged with ensuring that the text to be printed was not owned by another printer or bookseller. Printers who were not members of the Guild, in particular those operating outside of London, were not however subject to the powers of the Wardens or the Guild. This was a particular concern for both the Church and the Crown. With the proliferation of printed matter, literacy rates within the general population grew rapidly and within the literate populous demand for satirical material was high. To meet this demand many publications were produced which were highly satirical of the Crown while others attacked the Church. As a result, the world's first press regulations were promulgated in a direct attempt to control these publications.[38]

By a proclamation of June 1530, the Church was given a pre-emptive censorship over theological books. The proclamation ordered that new theological books in English were not to be sent to press before they had been examined by the Bishop of the local diocese.[39] Similarly, by a further proclamation of 1538 no secular book in English was to be printed in England without the approval of a Royal Licenser.[40] Thus by 1540 two distinct regulatory regimes had been developed to deal with the threat posed by the printing press. Controls designed to regulate competition had been developed by the Guild, while a licensing system had been developed by the Church and Crown to control content. The two regimes coincided through their enforcement mechanism, as it was the Wardens of the Mayor of London who both approved a book for registration within the Register of the Guild and who ensured any new book had the necessary ecclesiastical or civil authorisation. It is these two distinct aims: to protect markets and to regulate potentially seditious material, which went on to form the foundations of modern copyright law. At this time the requirements of the Guild and the Crown and Church coincided in a desire to exclusively control the printed word: a type of control to be found in exclusive property rights. Over time the

---

36 Blagden, C, *The Stationers' Company: A History 1403–1959*, 1960, London: Allen, pp 22–3.
37 Ibid, pp 32–3.
38 Note here the term press is being used in its correct meaning – being a machine for the printing of text.
39 Blagden, above fn 36, p 30.   40 Ibid.

common desires of both sets of parties drew them closer together and in 1557 Mary Tudor granted to the Stationers' Company (as the Guild had become known) a Royal Charter – in so doing effectively appointed them as agents of the Crown. The preamble of the Charter declares that the King and Queen, wishing to provide a suitable remedy against the seditious and heretical books that were daily printed and published, gave certain privileges to their beloved and faithful lieges, the 97 Stationers.[41] The Charter's key provision ordained that:

> no person within this our realm of England or the dominions of the same shall practise or exercise by himself or by his ministers, his servants or by any other person the art or mistery of printing any book or any thing for sale or traffic within this our realm of England or the dominions of the same, unless the same person at the time of his foresaid printing is or shall be one of the community of the foresaid mistery or art of Stationery of the foresaid City, or has therefore licence of us, or the heirs or successors of us the foresaid Queen by the letters patent of us or the heirs or successors of us the foresaid Queen.[42]

The Charter thus created a legal monopoly over publishing and vested this monopoly in the Stationers' Company. This is the first time we find exclusive legal control over creative goods being vested in a single entity within any jurisdiction. By this action the Crown created a quasi-proprietary right over the creation of printed texts: this is effectively the world's first copyright provision. To ensure efficient enforcement of this provision, the Master, Keepers or Wardens of the Company were given the right to search the houses and business premises of all printers, bookbinders and booksellers in the Kingdom for any printed matter, to seize and treat as they thought fit anything printed contrary to any statute or proclamation, and to imprison anyone who printed without the proper qualification or resisted their search.[43] The artificial propertisation of printed texts was thus brought about by a desire on the part of the publishers to protect their market position against the challenge posed by free riders, and by a desire on the part of the Crown to control the content of publications.

---

41 The preamble of the Stationers' Company charter reads: 'Know ye that we, considering and manifestly perceiving that certain seditious and heretical books rhymes and treatises are daily published and printed by divers scandalous malicious schismatical and heretical persons, not only moving our subjects and lieges to sedition and disobedience against us, our crown and dignity, but also to renew and move very great and detestable heresies against the faith and sound catholic doctrine of Holy Mother Church, and wishing to provide a suitable remedy in this behalf'. Full text available at: *http://victoria.tc.ca/~tgodwin/duncanweb/documents/ stationers_charter.html*.

42 Ibid.   43 Ibid.

Today most textbooks on intellectual property begin their examination of the law of copyright with the much later Statute of Anne.[44] Few though explain the background to the Act's promulgation. The Act, as we can see, did not develop in a regulatory vacuum: it was promulgated, at least in part, to frustrate extensive abuse of the publisher's copyright, which had been aggressively developed by the Stationers Company. In 1583, a Royal Commission criticised the existence of the right;[45] however, the response from the Crown was to further strengthen the power of the Company, secure in the knowledge that it was doing an excellent job of suppressing seditious comment.[46] The Company continued in this vital role for successive Tudor and Stuart monarchs throughout the remainder of the sixteenth and the seventeenth centuries. As a result, the Company developed an extensive power base, with successive monarchs taking steps to secure the Company's independence and enforcement rights – the common aims of each party to this Faustian pact ensuring that the copyrights of the members of the Company were prosecuted to the fullest extent of the law. It was only following the Glorious Revolution of 1688 that the power of the Stationers' Company fell into decline. Finally, in 1695, its power seemed to be extinguished when the Printing Licensing Act of 1642 lapsed: for the first time since 1557 there was no copyright in England and no one had the power to search for or seize unlicensed or illegal books. It was this extinction of the traditional publisher's right which led to calls from the Stationers' Company and its members for a new 'copyright act'. Between 1695 and 1707, the Stationers' Company made 10 unsuccessful attempts to lobby for legislation to restore the Licensing Act or for a new scheme for the registration of copyright.[47] Their attempts though were ultimately to end in failure when on 10 April 1710 the Statute of Anne came into force. The Act, instead of restoring the traditional publisher's copyright, created a new right of 'copyright', which removed ownership of creative goods from the publishers and vested it instead in the author – in effect rewarding the creative talent rather than the mercantile talent. In addition, the new Act enacted a series of provisions designed to alleviate the potentially deleterious market effects that the recognition of such a monopoly right may have, including, most importantly, the introduction of a time limit, or term, set at 14 years from the date of first publication – a significant departure from the Stationers' traditional practice

44 Holyoak, J and Torremans, P, in their text *Intellectual Property Law*, 4th edn, 2005, Oxford: OUP state after a short discussion of the Stationers' Company that the Statute of Anne is 'the first real copyright statute' (at p 9); Bentley, L and Sherman, B, in their text *Intellectual Property Law*, 2nd edn, 2001, Oxford: OUP record that 'most histories of British copyright law tend to focus on the origins of copyright, which are usually traced back to the 1710 Statute of Anne' (at p 29).
45 Blagden, see above, fn 36, p 42.   46 Ibid, pp 71–2.
47 Saunders, D, *Authorship and Copyright*, 1992, London: Routledge, p 51.

of a perpetual copyright. In effect, it abrogated the Stationers' monopolies at the stroke of a pen. This vital contextualisation of the Statute of Anne is missing from those textbooks that take up the story with the passing of the Act. Although it is correct to say that the Statute of Anne was the first Act to introduce the concept of the author's copyright and the concept of a copyright term, when drafted the Act was designed not only to usher in the new regime, but equally to sweep away the last vestiges of the earlier publisher's copyright system.

An appreciation of the earlier publishers copyright system is essential as it is that which provides the roots of our modern system of copyright. It is because the Church, the Crown and the publishers shared a common need to control the market in printed text that the concept of a proprietary right in creative works developed. Further, it was through the pact between the Crown and the Stationers' Company that this grew to become a powerful and commercially significant right. The role played by the Stationers' Company in the original development of the copyright also explains why the proprietary right attaches to the physical carrier of the text rather than to the text itself. In part, we protect books (the bottle rather than the wine) for reasons of convenience: printing 'fixes' a text in a definitive form and in cases of counterfeiting, the process of printing is highly visible and a printer is usually easy to identify. But in part, quite a substantial part, we protect books because the Guild of Writers of Text-letters, Lymners, Bookbinders and Booksellers took steps in the early sixteenth century to prevent free-riding in the *publishing* industry through the creation of a quasi-proprietary publisher's copyright in printed texts, a development that has influenced the course of copyright and copyright law ever since, but which now finds itself under intense scrutiny today for the first time since the 1440s.

Selling wine without bottles on the global net[48]

Digitisation is causing the divorce of content and carrier: there is no need to 'fix' content in the digital environment and no role for the traditional publisher. Now, as noted by John Perry Barlow, 'with the advent of digitization, it is now possible to replace all previous information storage forms with complex and highly liquid patterns of ones and zeros'.[49] This process of 'unfixing' is causing disintermediation within the market for creative goods, leading to the largest market upheaval since Gutenberg devised his press. The response from the media distribution industries – that is the collective might of book and magazine publishers, music producers, film studios, television broadcasters, radio broadcasters and newspaper publishers

---

48 This section heading is also taken from John Perry Barlow's classic article, 'The Economy of Ideas: Selling Wine Without Bottles on the Global Net', see above, fn 29.

49 Ibid.

– has been unfortunately predictable. The creation of a proprietary right in creative goods was, as we have seen, not a measured and balanced development designed to encourage and reward artistic creativity which, thanks to the Renaissance, was flowering across Europe; instead, it grew from protectionist measures nurtured by one of the strongest cartels of the Tudor period. It is therefore no surprise to find that as digitisation and disintermediation upset the careful equilibriums that the media distribution industries have nurtured within the market for creative goods, their first response was to seek to retrofit the digital environment with the design and hierarchy protections they had come to rely upon in the real world. The initial response from the media distribution industries was therefore an attempt to re-effect their exclusive control over creative content by recreating the market conditions that could be found in the real world. This involved two distinct, but related approaches: first, to attempt to refix creative content through the use of a design/hierarchy hybrid technique called Digital Rights Management (DRM); and second, to re-establish their exclusive legal rights over such content through the extension of traditional copyright laws into cyberspace.

The first stage of this process involved the application of code to refix creative content to a carrier medium. The system developed to fulfil this requirement entails a complex mix of licences and encryption technology and is known as Digital Rights Management or more commonly just DRM. The core of the DRM system is the use of *ex ante* licences. Instead of buying the content in a fixed carrier medium, the consumer purchases a licence granting certain rights of access and use. The licence is a digital data file that specifies what rights of use the consumer has in relation to the content transferred. Usage rules can be defined by a range of criteria, such as frequency of access, expiration date, restriction of transfer to other devices and permissions. These rules can be combined to enforce certain business models, such as rental or subscription, try-before-buy, pay-per-use and many more. The licence conditions are enforced through design controls, specifically through data encryption techniques. The content itself is contained in a separate module, called quite logically the content module, which is locked using a digital encryption system.[50] The encryption system used in DRM modules is actually quite complex and involves a two-stage encryption process, which was described in some detail by Paul Ganley in his paper *Access to the Individual: Digital Rights Management Systems and the Intersection of Informational and Decisional Privacy Interests*:

---

50 For more on the mechanics of DRMs see Ganley, P, Access to the Individual: Digital Rights Management Systems and the Intersection of Informational and Decisional Privacy Interests, 2002, 10 *International Journal of Law and Information Technology*, 241, pp 243–6.

Two methods of encryption are utilised in a DRM. Content itself, such as an audio or text file, is rendered unintelligible using a method called *single key encryption*. The content in this form is encrypted in such a manner as to make it impossible to use unless the same encryption key is applied to the content to return it to its original form. The weak link in this method is having to transport the key itself from the supplier to the end-user. *Public Key Encryption* provides a solution. By separating a key into two parts – one private, one public, *both* of which are needed to perform the encryption/decryption sequence – the supplier can encrypt a data packet containing both the encrypted content and the key itself using an end user's public key, who upon receipt can decrypt the data packet using their own private key which no-one has access to.[51]

This system of double encryption sounds both exceedingly complex and exceedingly secure, when in truth it is really neither. The system is quite simple and is widely used by, among other things, the Secure Socket Layer (SSL) technology, which is employed by internet browsers to protect sensitive data such as financial records in the online environment. The system uses two separate encryption protocols – one a single key encryption protocol, the other a public key protocol – for reasons of cost, convenience and robustness. Single-key encryption, where a matched set of encryption keys is used to both encrypt and decrypt the data is cheaper and simpler to implement and can be made more robust for less investment. It is therefore ideal for low-cost transactions such as the online purchase of music, text, video or other content. The weakness of single-key encryption is that the decryption key must be transferred from the supplier to the end-user. This makes the key vulnerable to interception, undermining the robustness of the encryption protocol. The remedy is to send the encryption/decryption key in an encrypted format. This is where public-key encryption (PKE) is valuable. In PKE, two keys are created that are mathematically linked, but which are undiscoverable from the properties of each other. One is kept private, the other is published for all to see. When the supplier of DRM encrypted material wishes to transmit the content to the consumer, s/he uses the consumer's public key[52] to encrypt the decryption key before sending it to the consumer. Upon receipt the consumer uses his/her private key to decrypt the decryption key, which then in turn is used to decrypt the content module. You may question why DRM systems use both encryption protocols: why not simply encrypt the entire content module using PKE? The reason is that PKE is considerably more complex and costly to employ in comparison with single-key encryption; it would simply be

---

51 See Ganley, above, fn 50, p 244.
52 The consumer will have a public/private keypair, which will have been generated by the DRM software when installed on his computer.

uneconomic to encrypt low-value content such as a music file by PKE as every content module transmitted would have to be individually encrypted using each individual consumer's public key. It is much easier to encrypt the content module using a standard encryption protocol, then simply transfer that key subject to the conditions of the licence. Through this complex web of licence conditions and encryption techniques, suppliers of creative goods attempt to digitally recreate the physical act of fixing creative content. Much like attaching it to the page, the content becomes fixed by being attached to the content module. Thus for the media distribution industries, DRMs offer an opportunity to replicate the physical protection offered by carrier media and as a result they have been widely adopted in systems such as the Content Scramble System used on DVDs,[53] Cactus Data Shield used on audio CDs,[54] Apple's 'Fair Play' system used on iTunes[55] and Sony's ATRAC standard, which is used by all Sony digital audio devices.[56] Despite their widespread use, DRMs are in many ways quite unlike traditional carrier media. The ability of a DRM to 'lock' content, thus preventing access – not merely copying – means that they are among the most contentious areas of technology.

Supporters of the Creative Commons frequently point out that DRMs are more invasive than traditional methods of fixing creative works. Whereas text fixed on paper could always be accessed by the end-user and could be copied, albeit with some degree of difficulty, in accordance with legal permissions such as the fair dealing provisions found in the Copyright, Designs and Patents Act,[57] text encrypted by a DRM system cannot be accessed unless one complies with the necessary provisions of the licence module, including payment if required, and copying will be strictly managed, perhaps even blocked, without reference to statutory entitlements. Thus as noted by Heather Ford, the organiser of Creative Commons South Africa:

> The main problem is that technological protection measures (like Digital Rights Management) often don't allow for the exceptions and limitations on copyright that enable us to exercise our fair use/dealing rights. Making copies of a book for personal research purposes, for example, or transporting texts into formats for the blind and differently-abled are actions that we take for granted in the analogue world, but make for huge problems when we want to do the same things on the Net.[58]

---

53 CSS is one of the oldest DRM systems and was introduced in 1996 by the DVD Copy Control Association, an industry body made up of leading hardware suppliers and content providers.
54 Cactus Data Shield (CDS) is produced by Macrovision, the leading supplier of DRM technology. For more on CDS see *www.macrovision.com/products/cds/index.shtml*.
55 For more on 'Fair Play' see *www.apple.com/lu/support/itunes/authorization.html*.
56 For more on ATRAC see *www.minidisc.org/aes_atrac.html*.
57 Copyright, Designs & Patents Act 1998, Chapter III.
58 Ford, H, *Should Digital be Different?* 2005, available at: *http://za.creativecommons.org/blog/archives/2005/02/25/should-digital-be-different/*.

This result is caused by the convergence of two developments, and is of particular concern not only to supporters of the Creative Commons, but to anyone who consumes creative goods. The first cause of this is the closing of the so-called analogue window. The analogue window is caused by the disruptive effects of the real world – in Nicholas Negroponte's term, the world of atoms. The atomic world is currently too complex to allow for complete protection of physical works, however fixed, as at some point they are turned into waves (either light or sound), which may be intercepted and replicated. Thus a book may always be photocopied (a technique that uses reflected light waves from the original to make copies), or a music recording may always be recorded using a microphone and a recording medium. There is no effective way to prevent analogue copying of this nature, but fortunately for the producers of creative products, Mother Nature has provided a kind of analogue protection through 'drop-off' or loss in quality in successive generations of copy. Thus in the analogue world an uneasy truce had emerged between suppliers and consumers of creative goods. The suppliers knew that the consumers would make photocopies of books and texts (how many of you are now reading this on a photocopy?); that they would make illegal tape recordings of records and CDs or from radio broadcasts and that they would use video tapes to record movies and dramas from television to retain permanently.[59] The result of this was a market response wherein those of us who bought original materials paid a higher price (a kind of informal pirate tax) to subsidise the losses caused by home recording.[60] The introduction of digital media has substantially affected this delicate settlement. The digital world is built from bits, not atoms, and this 'shift from atoms to bits'[61] has encouraged suppliers to attempt to take control of their product throughout the supply chain.

As creative products (with the exception of printed material) are no longer shipped in analogue format, the analogue window has closed.[62] The closure of the analogue window provides the foundations of our second development. The ability to control digital media products in all parts of the supply chain means that the suppliers of digital media may effect direct control on the

---

59 Although s 70 of the Copyright, Designs & Patents Act 1998 allows 'The making for private and domestic use of a recording of a broadcast or cable programme solely for the purpose of enabling it to be viewed or listened to at a more convenient time', it does not allow for this copy to be kept permanently for repeated viewing.

60 In most countries the settlement was more apparent with a blank media tax being imposed, which would, at least in part, be allocated to content providers. Countries that impose (or imposed) such a tax include Canada, the US, Sweden, France and Germany. The UK is unusual among western nations in not having such a levy.

61 Negroponte, N, *Being Digital*, 1995, Chatham: Hodder & Stoughton. Ch 1.

62 Purists would point out that there remains a slight analogue window that can probably never be eliminated and this is the final conversion of the content to light or sound waves for detection by our eyes/ears. Thus I can always take a video camera and record the image being displayed on my TV screen, although it should be noted that Macrovision supplies DRM tools to guard against this eventuality as well.

end-user through the contractual terms contained in the DRM licence module. In effect this sees another fundamental shift in the established settlement, this time from regulation by public authorities to regulation by private entities. The resultant effect is therefore a quasi-privatisation of copyright law, for as noted by Dan Burk and Julie Cohen, 'The design of technological constraints, however, is not the sole provenance of the state; indeed, it is more often left to private parties. In the case of rights management systems, copyright owners determine the rules that are embedded into the technological controls'.[63] It should be noted though that these rules do not and cannot legally supercede the established public provisions such as are found in Chapter III of the Copyright, Designs and Patents Act. But, there is a key difference between what is legally possible and what is effectively possible and Burk and Cohen note this:

> By implementing technical constraints on access to and use of digital information, a copyright owner can *effectively* supersede the rules of intellectual property law ... The implications of this development are stark: Where technological constraints substitute for legal constraints, control over the design of information rights is shifted into the hands of private parties, who may or may not honor the public policies that animate public access doctrines such as fair use. Rights-holders can effectively write their own intellectual property statute in computer code.[64]

The key word in this extract is 'effectively'. Although legally a copyright owner cannot restrict access to material in accordance with fair use/fair dealing provisions, they can control how and when such access occurs. For example if I wished to use an excerpt from an episode of *The Simpsons* to illustrate a point in a lecture, then legally I could make use of this excerpt without the permission of the copyright holder under s 34 of the Copyright, Designs and Patents Act. There is no legal or actual impediment to prevent me from so doing. Now imagine that my copy of *The Simpsons* is a digital file protected by DRM technology and delivered to my computer. I decide I wish to use an excerpt from the copy for teaching purposes, but the licence restricts use of the file to my computer, and restricts viewing access to the complete file (that is I cannot excerpt it). I still have the legal right to use the file as a teaching tool as I had previously, but there is now a physical impediment that prevents me from so doing. I must now take positive steps to contact the copyright holder and to ask for them to supply to me the excerpt I require in a format that will allow me to use it. Now I am required to take positive action to enforce my right, this increases my opportunity cost of so doing, making it

---

63 Burk, D and Cohen, J, 'Fair Use Infrastructure for Copyright Management Systems', 2001, 15 *Harvard Journal of Law & Technology*, 41, p 49.
64 Ibid. Emphasis added.

less likely that I will enforce my right. The shift from atoms to bits, rather than making it easier for the consumer to exercise his/her legal right has in fact led to a narrowing of choice. It is this that causes the effective control referred to by Burk and Cohen.

(Un)fair competition in the market for digital products

It would appear that the suppliers of creative goods in the digital environment had achieved complete control over their products and their markets. Whereas arch-cyber-libertarian John Perry Barlow had predicted that digitisation would set information free,[65] it now appeared that by *fixing* creative goods through DRM technology the media distribution industries have achieved an unassailable control over such goods. Through a bold manoeuvre they had not only managed to retain the control over informational products that they had held in the analogue market, but they now had the opportunity to control such products when they left their possession – control that had been lost with the invention of the photocopier, the microcassette recorder and the video recorder. In fact such is the level of control that DRM technology offers that arguably it gives suppliers the most complete protection over creative works since the invention of the Gutenberg Press. But the potential to exercise such control is one thing, it is quite another to actually achieve it. If we were to assume that the introduction of DRM technology led to the conclusion suggested above, we would have learned nothing from the tale of the Gardener's Dilemma featured in Chapter 2. The consumer market for creative goods did not remain static while the suppliers developed and perfected their tools of control, and the tumultuous developments that occurred in the consumer marketplace during this time undermined the effectiveness of DRM technology.

The first key development was in the technology used to supply informational products. When DRMs were first proposed in the early 1990s it was envisaged that digital informational products would be distributed primarily through traditional channels such as high-street stores and broadcast outlets. With the explosive success of the web, this model was changed to acknowledge the possibility of the employing of online distribution channels. This revised model though retained the assumption that creative goods such as music and film would still be downloaded from a variety of centralised

---

65 In his classic article, 'The Economy of Ideas: Selling Wine Without Bottles on the Global Net', see above, fn 29, Barlow stated: 'all the goods of the Information Age – all of the expressions once contained in books or film strips or records or newsletters – will exist either as pure thought or something very much like thought: voltage conditions darting around the Net at the speed of light, in conditions which one might behold in effect, as glowing pixels or transmitted sounds, but never touch or claim to "own" in the old sense of the word.'

(commercial) servers: the model now used by systems such as Apple's iTunes service. Further, it was assumed that with these centralised distribution points controlling the supply of content, end-users would happily comply with DRM conditions, benefiting all parties by allowing for a much finer pricing and licensing model than traditional distribution media allowed. Thus music could be sold: on a pay-to-play basis (where you pay a small fee for each use); on a subscription basis (as is being used now by Napster's *Napster To Go* service); as an outright sale (as used by iTunes), or any variety or combination of such permissions. This meant that if you were the kind of user who listened to music for a few months, then moved on, you didn't have to pay to buy the full licence – an option not possible at the time, due to the limitations of distributing music on physical media such as CDs or tapes. In fact, when DRMs were eventually introduced into the music market a variety of factors conspired to produce an unexpected consumer response. Technological failures caused consumers to initially reject DRM technology affixed to traditional supply media such as CDs, as can be seen from the Bertelsmann example. In autumn 2001, Bertelsmann Music Group (BMG) fixed the Cactus Data Shield (CDS) DRM system to some of their new CD releases. The system, though, proved to be too complex for older home audio equipment and many consumers found that the protected CD simply would not play on their existing audio systems. This led to such high levels of consumer complaints that the company was forced to issue 'clean' copies of the all the CDs in question. It took over a year for BMG to return to the market with an updated CDS protocol, and although this system seemed to alleviate most of the problems, allowing the music industry to flex its market muscle to force the issue of DRM-protected CDs, the matter was soon to become unimportant as the music market had, in the interim, undergone a dramatic change in the way we collect, access and manage our music collections. The change was driven by two sources: one legal, one illegal. The legal contribution came from Apple Computers (and their imitators), which revolutionised the music industry when, on 23 October 2001, they launched the iPod music player and associated iTunes software. With the launch of these two products, Apple successfully merged the carrier medium with the player removing the need for music to be stored or transported in a separate physical media.[66] With the success of the iPod the potential market for digital media files exploded,[67] but

---

66 It should be noted Apple did not produce the first digital audio player; this honour went to Eiger Labs with their Eiger Labs MPMan F10, which was launched in the summer of 1998. With its limited 32MB memory though it could only store a few tracks. Apple's high volume 5GB iPod could transport around 1,000 songs and through clever design and marketing created the market for high-capacity digital audio players.

67 Latest estimates suggest that iTunes has downloaded over 750m tracks since its launch, see *www.theregister.co.uk/2005/09/20/apple_jobs_piracy_pricing/*, while figures from the International Federation of the Phonographic Industry reveal that the level of legal downloads

the traditional media distribution industries were caught out by the success of the iPod. They had not developed their digital distribution systems and were therefore not able to exploit the new market opportunity offered by music downloading. This meant that the market opportunities this offered were about to be taken up by someone else, and with Apple's iTunes music store charging 99¢ per track in the US and 79p per track in the UK it was clear that someone would undercut them in the marketplace.

The source of this competition had become apparent in the autumn of 1999 when a student at Boston's Northeastern University, called Shawn Fanning, released his 'Napster' protocol. Fanning created Napster out of frustration. He was, as many college students are, an avid music fan who was strapped for cash. He was frustrated for several reasons: first, he wanted to search for digital music files, but the only option available at the time was to use crude search engines that would search the entirety of a library with no specific ability to search for music files;[68] second, he wanted to swap interesting pieces of music with like-minded individuals, but didn't have the tools to do so; and third, he, like many others at that time, was frustrated by the quality of music available and the cost of replacing older collections on vinyl with newer collections on digital media.[69] Fanning designed his Napster protocol to meet these needs. It was, in his mind at least, primarily a tool designed to create a community where people could meet and talk about music. He initially envisaged that any trading of music files would take place outside the Napster community by email or IRC, and it was almost as an afterthought he added what turned out to be Napster's killer application. Fanning decided late in the development of the Napster protocol to add a revolutionary option within the protocol: the ability to interface directly with the computer of another Napster user and to download from his or her PC, music files in MP3 format. This concept – one of a network made up of thousands (or millions in the case of Napster) of computers, which may interact directly with each other without the need for their communications to be routed via a central server, known as peer-to-peer (P2P) or grid computing – was not new. It had been part of RFC-1, the first document to describe the ARPANET, but until Fanning built it into his Napster protocol no one had found a commercial use for it. As we now know, Napster was an instant success. It rapidly gathered members from around the globe, and although

rose by over 300 per cent from the first half of 2004 to the first half of 2005. See *www.ifpi.org/site-content/press/20050721.html*.

68 Systems in use at that time included Lycos for the web and IRC Search for Internet Relay Chat.

69 At the time Napster was released, there was a general perception that the quality of new albums had decreased. Many people said that albums contained only one or two good songs, along with many low-quality 'filler' songs.

the exact number of members it gained is not clear,[70] there is no doubt that Napster fundamentally altered the market structure for online music distribution with, at the peak of Napster's popularity, almost 3 billion music files being traded among members each month. In design, Napster was much like eBay: a consumer-to-consumer (C2C) trading community, but Napster was designed specifically around a single product – digital music files. This specificity meant that the way Napster functioned was quite unlike other C2C communities. These differences drove the early success of Napster, but were ultimately to lead to its downfall. As has been stated repeatedly throughout this chapter, creative goods are quite distinct from physical goods and whereas eBay is a valuable C2C reselling community that allows individuals to sell on items at the value the market attaches to them, members of the Napster community were engaged in something quite different. Napster was a C2C trading community – that much is true – but with Napster, the trading was in *copies* of music files, meaning the 'seller' never relinquished their original file. This in turn meant that there was no need to charge for files and so all music in the Napster community was available at no cost. The result was a market that was built on a clearly illegal activity and which fundamentally undermined the market model for paid-for digital music downloads: why pay Apple 99¢ per track when you could download it for free from Napster? The Napster market model was about to undermine the entire exercise of designing paid-for music download models. From the point of view of the media distribution industries it had to be closed down, and quickly.

As the vast majority of music available on Napster was protected by copyright, a group of leading music studios, including A&M Records, Geffen Records, MCA, Motown and Capitol Records, raised a suit against Napster claiming contributory and vicarious copyright infringement.[71] Following a trial hearing the District Court found that the complainants had successfully established a prima facie case of direct copyright infringement on the part of Napster's users. According to the Court, 'virtually all Napster users engage in the unauthorized downloading or up-loading of copyrighted music',[72] and 'Napster users get for free something they would ordinarily have to buy [which] suggests that they reap economic advantages from Napster use'.[73] Further, the

---

70 Napster CEO Hank Barry claimed that there were 'more than 60 million members of the Napster community', but PC Data Online recorded only 18.7 million users and Media Matrix put the figures for **active** users as low as 6.7 million See statistics at: *www.clickz.com/stats/sectors/retailing/article.php/501021.*

71 *A&M Records Inc v Napster Inc* 114 F Supp 2d 896 (N D Cal 2000), affd in part and revd in part, 239 F 3d 1004 (9th Cir 2001). Note: as the file transfer took place between customers, Napster was not involved in direct copying of protected works, *but* as Napster provided the searchable database of files, and as Napster initiated contact between customers the RIAA argued they were both contributorily and vicariously liable for the copyright infringement which took place.

72 *A&M Records Inc v Napster Inc* 114 F Supp 2d 896, p 911.   73 Ibid.

Court held that the effect of the use upon the value of the work and potential markets for the work weighed against finding that use of Napster constituted fair use. As a result it rejected Napster's fair use defence, and distinguished the Supreme Court's decision in *Sony Corp of America v Universal City Studios*.[74] In particular, the trial judge noted that unlike VCRs, in which users were initially invited to view the television broadcast for free, Napster users obtained permanent copies of songs that they would otherwise have had to purchase. Further, the majority of VCR users merely enjoyed the tapes at home; in contrast, 'a Napster user who downloads a copy of a song to her hard drive may make that song available to millions of other individuals . . . facilitat[ing] unauthorized distribution at an exponential rate'.[75] The District Court, in short, concluded that the conduct of Napster users could not be considered fair use because it threatened the incentives created by copyright. With this finding, the music industry obtained judgments against Napster for both contributory infringement,[76] and vicarious infringement.[77] An appeal by Napster to the Ninth Circuit proved to be unsuccessful[78] and in February 2001, Napster was closed down.[79] What the music studios failed to appreciate though was that although the litigation undertaken by them had succeeded in

---

74 464 US 417 (1984).   75 *A&M Records v Napster*, see above, fn 72, p 913.

76 Contributory infringement, requires both knowledge of the infringing activity and a material contribution (actual assistance or inducement) to the alleged primary infringement. The court interpreted the knowledge requirement as not merely that the Napster system allowed an infringing use, but that Napster had actual notice of the infringement and then failed to remove the offending material. The Court concluded that Napster knew or had reason to know of its users' infringement of plaintiffs' copyrights, that Napster failed to remove the material, and that Napster materially contributed to the infringing activity by providing the site and facilities for direct infringement.

77 Vicarious infringement results when there has been a direct infringement and the vicarious infringer is in a position to control the direct infringer, fails to do so and benefits financially from the infringement. The Court held that Napster was vicariously liable as they failed to exercise their right and ability to prevent the exchange of copyrighted material. Further, Napster had a direct financial interest in the downloading activities since their revenue was dependent on user increase which was driven by the infringing activities of users.

78 In fact Napster's appeal was partly successful as the Court of Appeal noted that: 'contributory liability may potentially be imposed only to the extent that Napster: (1) receives reasonable knowledge of specific infringing files with copyrighted musical compositions and sound recordings; (2) knows or should know that such files are available on the Napster system; and (3) fails to act to prevent viral distribution of the works. The mere existence of the Napster system, absent actual notice and Napster's demonstrated failure to remove the offending material, is insufficient to impose contributory liability.' (at p 1014). This meant the plaintiff's had to give Napster written notice of all infringing files.

79 On its last day of full operation the 10 most popular downloads on Napster were: (10) Doobie Brothers: Listen to the Music; (9) Red Hot Chili Peppers: Give it Away; (8) Dr. Dre: Bang Bang; (7) Metallica: Seek and Destroy; (6) Jimmy Buffet: A Pirate Looks at 40; (5) Warren Zevon: Send Lawyers, Guns and Money; (4) Judge Jules: Gatecrasher; (3) Jerky Boys: Fanning my Balls (probably a play on the founder's name); (2) The Clash: I Fought the Law (and the law won); and (1) Everly Brothers – Bye Bye Love.

closing the Napster file-sharing community, it had failed to contain the technology that underpinned it. Others had been watching the Napster litigation and had learned from it. Napster's downfall, it seemed, was the implementation of a centralised database of files. It was this that the court had found persuasive of the claim that Napster had knowledge of the illegal activities of its members. If you could engineer a P2P system without such a server then you would have a defence against Napster-style claims.

Two such systems were already in development when the Napster decision came out. The first was the Gnutella protocol developed by Justin Frankel and Tom Pepper of Nullsoft, and which was released, apparently in error, on 14 March 2000. Gnutella is a completely distributed P2P protocol, meaning that it had no central servers or databases. It can be used to transfer any type of digital file, but is most widely used for music and video files. It is supported by several clients, including BearShare, Limewire and iMesh. The second was the FastTrack protocol, which was developed by Niklas Zennström and Janus Friis and was introduced in March 2001 by their company Consumer Empowerment. FastTrack is a second-generation P2P protocol, which uses a partially decentralised network wherein some peers act as a *supernode*, a temporary indexing server for other, slower, clients. Like Gnutella, FastTrack can be used to transfer any digital file, but is most popular with users looking to swap music and video files. It is supported most commonly by the KaZaA client. With no central server to attack in either of these protocols, or in the later BitTorrent and eDonkey P2P protocols, it proved impossible for the copyright owners to deal a single deadly blow as they had done with Napster. The media distribution industries now faced a new, and more acute, challenge. Developers of P2P systems had learned from Napster's downfall. With no central server they had no knowledge of the use their protocol was being put to. Further, with the removal of the central database it also meant there was no weak point in the system to attack: P2P now functioned more like the internet itself. Not only was the software now more robust, but so too were the commercial structures that supported and distributed it. When the media distribution industries embarked upon their protracted litigation against the providers of second-generation P2P services in late 2002, they found, for instance, that just one such provider, Sharman Networks, provider of the KaZaA client, was 'incorporated in the South Pacific island nation of Vanuatu and managed from Australia. Its computer servers are in Denmark and the source code for its software was last seen in Estonia while KaZaA's original developers, who still control the underlying technology, are thought to be living in the Netherlands, although entertainment lawyers seeking to have them charged with violating US copyright law have been unable to find them'.[80] Due to this

---

80 Harmon, A, 'Music Industry in Global Fight on Web Copies', *The New York Times*, 7 October 2002, available at:*http://uts.cc.utexas.edu/~lcp/articles/music_industry_ asks_who_controlls_web.htm.*

complex corporate structure, litigation against KaZaA has already been ongoing for five years and it is still in progress.[81] The first round of litigation took place in November 2001 when the Amsterdam District Court gave KaZaA two weeks to cease infringing recording artists' copyrights, following which it would face a penalty of 100,000 guilders a day.[82] KaZaA's immediate response to the judgment was twofold. First, it appealed the judgment, and second, it sold its technology to Sharman Networks, taking it beyond the direct control of the Dutch legal system, whatever the outcome of the appeals process.[83] While continuing to prosecute the litigation in the Netherlands the media distribution industries turned their attention to KaZaA's primary market and its new owners. KaZaA was named, alongside several other file-sharing services, in a claim taken out by MGM and others in Los Angeles in October 2001. The complaint claimed that the 'Defendants are building a business based on the daily massive infringement that they enable and encourage,'[84] and sought a permanent injunction enjoining the defendants and defendants' agents, from: '(a) directly or indirectly infringing in any manner any of Plaintiffs' respective copyrights or other exclusive rights, and (b) causing, contributing to, enabling, facilitating, or participating in the infringement of any of Plaintiffs' respective copyrights or other exclusive rights.'[85] As Sharman Networks, the new owners of the KaAzA protocol were based outside the US; this was then followed by a subsequent action in Australia.[86] By good fortune important milestones were reached in both these cases in the summer of 2005. The US case finally reached the Supreme Court. By this time, though, KaZaA BV (the original Dutch Company) had gone into liquidation and attempts by the complainants to cite Sharman Networks as respondents had been unsuccessful, leaving only two respondents: Grokster Ltd. and StreamCast Networks Inc.. The respondents went into the Supreme Court hearing feeling confident having won both District Court and Appeals Court hearings, but the Supreme Court was about to issue

---

81 At the time of writing in May 2006.
82 *Buma/Stemra v KaZaA*, Amsterdam District Court, 29 November 2001, rolnummer KG 01/ 2264, available at: *www.solv.nl/rechtspraak_docs/KaZaA%20v.%20Buma%20Stemra%20% 20Pres.%20Rb.%2029%20november%202001.pdf*.
83 Interestingly the KaZaA appeal turned out to be successful with both the Amsterdam Court of Appeal and the Supreme Court of the Netherlands ruling that KaZaA could not be held liable for copyright infringement of music or movies swapped on its software. See *www.eff.org/IP/P2P/BUMA_v_Kazaa/20020328_kazaa_appeal_judgment.html* (Amsterdam Appellate Court) and *www.solv.nl/rechtspraak_docs/KaZaA%20v.%20Buma%20Stemra%20 %20Supreme%20Court%2019%20December%202003.pdf* (Supreme Court). These results have led several commentators to note that 'KaZaA may have yielded Dutch territory too soon'.
84 Taken from Complaint filed on 2 October 2001, available at: *www.eff.org/IP/P2P/MGM_ v_Grokster/20011002_mgm_v_grokster_complaint.pdf*.
85 Ibid.
86 *Universal Music Australia Pty Ltd v Sharman License Holdings Ltd* [2005] FCA 1242.

a landmark decision. In finding against the P2P providers, the Justices of the Supreme Court stated that the 'respondents unlawful objective is unmistakable . . . each of the respondents showed itself to be aiming to satisfy a known source of demand for copyright infringement, the market comprising former Napster users, with . . . neither respondent attempt[ing] to develop filtering tools or other mechanisms to diminish the infringing activity using their software'.[87] The decision was clear – the deliberate use of decentralised networks as a tool to mask the activities of customers from the view of service providers will not absolve that service provider from liability if it is clear they aim to profit from that activity and do nothing to prevent it. In so deciding the Supreme Court borrowed a notion from patent law known as the 'active inducement theory'.[88] The case was remanded back to the Court of Appeal for the Ninth Circuit for further proceedings, but the clear guidance of the Supreme Court was enough to convince most P2P network providers that they needed to settle the issue. Grokster announced in November 2005 that it would no longer offer its P2P file-sharing service. The notice placed on their website reads:

> The United States Supreme Court unanimously confirmed that using this service to trade copyrighted material is illegal. Copying copyrighted motion picture and music files using unauthorized peer-to-peer services is illegal and is prosecuted by copyright owners.
>
> There are legal services for downloading music and movies. This service is not one of them.
>
> YOUR IP ADDRESS IS 158.143.69.87 AND HAS BEEN LOGGED. Don't think you can't get caught. You are not anonymous.

In addition to closing their service Grokster paid damages said to amount to $50 million to the music and recording industries. Another popular P2P client, BearShare, agreed a settlement in May 2006, which saw it pay damages amounting to a reported $30 million in settlement of copyright infringement claims.[89] Owners and operators of P2P software suites have even appeared before Congress to announce that they will work with the music industry to free their networks of infringing material.[90] Despite this new dawn of

---

87 As *MGM et al v Grokster et al* 125 S.Ct. 2764 (2005). Slip opinion available at: *www.eff.org/ IP/P2P/MGM_v_Grokster/04–480.pdf*

88 In Patent Law active inducers of patent infringement cannot escape liability by showing that they are selling a technology suitable for non-infringing uses.

89 See *www.foxnews.com/story/0,2933,194412,00.html*.

90 See testimony of Sam Yagan, President, MetaMachine Inc (developer of eDonkey and Overnet) before the United States Senate Commission on Judiciary, 28 September 2005, available at: *http://judiciary.senate.gov/testimony.cfm?id=1624&wit_id=4689*.

co-operation between the P2P industry and the recording industry, not all P2P providers are filled with such enthusiasm for co-operative efforts. StreamCast Networks, the last remaining co-defendant in the MGM litigation, filed a Response in Opposition to the entertainment industries' motion for summary judgment on 14 April 2006. In a press release they gave their reason for doing so: 'StreamCast's position that there are issues remaining to be decided by a jury at the trial court level – that is a jury needs to decide what the true facts are – since there are disputes between the parties as to what the true facts are and whether StreamCast intended to and did actively induce copyright infringement as evidenced by the true facts.'[91] Another provider still battling the recording industry is Sharman Networks. The ongoing litigation in Australia in the case of *Universal Music Australia v Sharman* remains of some import. On 5 September 2005 judgment was given in that case. In his judgment Mr. Justice Wilcox recorded that:

> despite the fact that the Kazaa website contains warnings against the sharing of copyright files, and an end user licence agreement under which users are made to agree not to infringe copyright, it has long been obvious that those measures are ineffective to prevent, or even substantially to curtail, copyright infringements by users. The respondents have long known that the Kazaa system is widely used for the sharing of copyright files.
>
> There are technical measures that would enable the respondents to curtail – although probably not totally to prevent – the sharing of copyright files. The respondents have not taken any action to implement those measures. It would be against their financial interest to do so. It is in the respondents' financial interest to maximise, not to minimise, music file-sharing. Advertising provides the bulk of the revenue earned by the Kazaa system, which revenue is shared between Sharman Networks and Altnet.[92]

This he said meant that Sharman had authorised infringement of copyright by users of its file-sharing software and he ordered Sharman to adopt technical measures to curtail infringement of copyright.[93] Sharman have appealed this decision, but where does this leave all the interested parties in the market for digital creative products: the suppliers, the network providers and the consumers?

---

91 See StreamCast press release at: *www.streamcastnetworks.com/vs.html*.
92 *Universal Music v Sharman License Holdings*, see above, fn 86.
93 A notice on the Kazaa website reads: 'To comply with orders of the Federal Court of Australia, pending an appeal, use of Kazaa Media Desktop is not permitted by persons in Australia. If you are in Australia you must not download or use the Kazaa Media Desktop.'

The media distribution industries obviously applauded both decisions and argued that they signalled the beginning of the end for illegal file-sharers.[94] Many commentators, though, disagree. For example in her excellent paper *Legally Speaking: Did MGM Really Win the Grokster Case?*[95] Professor Pamela Samuelson notes that:

> MGM didn't really want to win *Grokster* on an active inducement theory. It has been so wary of this theory that it didn't actively pursue the theory in the lower courts. What MGM really wanted in *Grokster* was for the Supreme Court to overturn or radically reinterpret the *Sony* decision and eliminate the safe harbor for technologies capable of substantial noninfringing uses. MGM thought that the Supreme Court would be so shocked by the exceptionally large volume of unauthorized up and downloading of copyrighted sound recordings and movies with the aid of p2p technologies, and so outraged by Grokster's advertising revenues – which rise as the volume of infringing uses goes up – that it would abandon the *Sony* safe harbor in favor of one of the much stricter rules MGM proposed to the Court. These stricter rules would have given MGM and other copyright industry groups much greater leverage in challenging disruptive technologies, such as p2p software. Viewed in this light, MGM actually lost the case for which it was fighting. The copyright industry's legal toolkit to challenge developers of p2p file-sharing technologies is only marginally greater now than before the Supreme Court decided the case.
>
> MGM is concerned that developers of p2p software will articulate a plausible substantial non-infringing use, such as downloading open source software, for their technologies and will be careful not to say anything that directly encourages infringing uses. MGM believes that they will nonetheless secretly intend to benefit from infringing uses that ensue. If there are no overt acts of inducement and no proof of specific intent to induce infringement, and if the *Sony* safe harbor continues to shield technology developers from contributory liability, MGM will find itself on the losing side of challenges to technology developers for infringing acts of their users. That is why MGM didn't really want to win the *Grokster* case on this theory.

---

94 Dan Glickman, President of the Motion Picture Association of America, said: 'Today's unanimous ruling is an historic victory for intellectual property in the digital age, and is good news for consumers, artists, innovation and lawful Internet businesses.', while John Kennedy, head of the International Federation of the Phonographic Industry said: 'It quite simply destroys the argument that peer-to-peer services bear no responsibility for illegal activities that take place on their networks.'

95 Available at: *www.sims.berkeley.edu/~pam/papers/CACM%20SCT%20decides%20MGM.pdf*.

Here, Samuelson identifies a vitally important concession made before the Supreme Court. Although the music industry was understandably keen to play up their somewhat unexpected success, the outcome is not exactly what it would have chosen before the decision was published. In particular, there remain two outstanding problems that the media distribution industries need to resolve before their success in the courtroom can be translated into success in the marketplace. The first is that this decision looks very much like a 'Napster II' scenario. While they have succeeded in obtaining an effective legal decision against a number of file-sharers, and although this will bring, and indeed already is bringing, these companies to the negotiating table, there remains in the background the next generation of P2P entrepreneurs awaiting the demise of Grokster, Limewire and StreamCast, before launching their services: services which no doubt will make no active inducement to copy.[96] Thus P2P networks are like the Hydra: no matter how many heads the media distribution industries cut off new ones will grow in the shape of new providers, using new technology, and who have learned from the fate of their predecessors. Second, as the KaZaA/Sharman Networks actions have shown, providers of P2P systems may use geography to shelter their activities. The decision in *MGM v Grokster* is geographically restricted to the confines of the US, and although it may signal a change in approach throughout western industrial nations,[97] some States, such as Vanuatu, will continue to host P2P providers at least in the medium term. Thus the reach of the *MGM v Grokster* decision is extremely limited, and despite somewhat bullish language it is as unlikely as the original *Napster* decision to curb file-sharing activities worldwide.[98] Taking legal action against the current generation of suppliers only deals with the immediate problem, not the underlying malaise. The only way to effectively curb the market in free (and illegal) copies of creative works is to shut off demand for such files: for as long as the demand remains buoyant, consumers will continue to trade files.

96 Some third-generation P2P protocols are already available. If a successor to Grokster were to distribute such software on a non-profit approach with the (claimed) view to it being used for the distribution of goods under the creative commons and free software licences it is hard to imagine how they could be claimed to be making an active inducement to copy.

97 The proposed *Directive on Criminal Measures Aimed at Ensuring the Enforcement of Intellectual Property Rights* COM(2005)276 final, would make supporting copyright infringement illegal across the EU.

98 In fact data from slyck.com reports that the number of active file-sharers on P2P networks has increased from 4.6 million users in January 2003 to 8.5 million users in May 2006, while Senator Dianne Feinstein noted in her opening statement to the United States Senate Commission on Judiciary hearing on 'Protecting Copyright and Innovation in a Post-Grokster World' that 'P2P user volume has continued to increase even after the decision'. Available at: *http://judiciary.senate.gov/member_statement.cfm?id=1624&wit_id=2626.*

Fixing digital products in indelible ink: encryption and control

Any hope the media distribution industries had of controlling the market for the distribution of digital creative goods is therefore lost. With the technological genie out of the bottle the P2P distribution model is now a major component of the market for digital goods. It might be expected at this point that the media distribution industries may take effective steps to recognise this reality and to design a strategy to contain the threat through the use of aggressive pricing of legal paid-for downloads and/or by offering a value-added service.[99] Although this is unlikely to 'kill-off' illegal file-sharing, it would help contain it to the same market sectors as illegal home taping: that is mostly children and young adults who cannot afford to pay for all the music they want to own and who have little regard for copyright rules. It appears though, that the industry has chosen not to enter into direct market competition with the file-sharers, but instead remains focused upon attempts to obtain monopoly control over the distribution of copyright protected digital products. In addition to the litigation against the P2P providers discussed above, this takes two forms: (1) to obtain complete control over the file itself through the application of DRM technology; and (2) the apparently illogical action of prosecuting its own customers.

As was discussed above, DRMs offer the media distribution industries the most tantalising opportunity: distribution of their products in a carrier medium that cannot be copied. As DRMs use an encryption key to control access to the media file they prevent not only illegal access but also copying – they are a bit like having a book printed on uncopyable paper where the text only becomes visible when the authorised owner of that book is reading it.[100] The problem with DRM encryption though is that any digital technology that can be engineered can also be reverse-engineered; in other words, months or years spent designing your encryption protocol may be undone in minutes by a cracker.[101] The question for the industry is how does it react to this threat?

An example of such a challenge, and the response of industry to it, may be seen in the DeCSS cases. These cases revolved around a weak encryption key

---

99 A recent report by In-Stat noted that 'While Peer-2-Peer and piracy issues have not entirely disappeared, consumers are showing heightened awareness and interest in legitimate online music services'. See In Stat, *Consumers' Willingness to Pay for Music Downloads Grows*, 18 April 2005.

100 It should be noted that DRMs do not on the whole prevent the digital media file from being replicated, but what is replicated is the *encrypted* file not the plaintext file, meaning that simply copying the file is worthless unless you have a licence from the copyright owner to access it.

101 Software cracking is the modification of software to remove encoded copy prevention. Those who carry out this activity are crackers, not hackers. The resultant decrypted file is know as Warez and are widely distributed.

called Content Scrambling System (CSS) introduced in 1996 to authenticate DVD discs and players. CSS was designed to ensure that pre-recorded DVDs could be played only on authorised players. It was simple: a set of matched keys including one key called the player key embedded in the DVD player itself. One particular problem with the CSS system was that there was no player key available for the Linux operating system (OS), meaning that Linux users could not view DVDs on their computer's DVD-ROM. The controllers of the CSS system, the DVD Copy Control Association, stated that this was due to the open nature of the Linux code, which meant that they could not protect against the illegal copying of content in Linux compatible format. Users of Linux suspected this was part of a campaign by the major content providers and the home electronics industry to 'kill off' the Linux platform and so they began to crack the CSS code. In October 1999, this project reached fruition when the CSS encryption code was successfully reverse-engineered. Popular myth suggests that this was done by a sixteen-year-old Norwegian youth called Jon Lech Johansen, although a text file attached to the cracked decryption key, called DeCSS, suggests it was actually produced by the Russian-based hacker/cracker community DrinkorDie, of which Jon Johansen was a member. Whoever actually engineered the DeCSS code, Jon Johansen took the credit, and the blame. The DVD Copy Control Association raised a civil action against him (and others) in the USA and he was arrested by Norwegian police and charged by the Norwegian National Authority for Investigation and Prosecution of Economic and Environmental Crime (ØKOKRIM). The criminal trial opened in the Oslo City Court on 9 December 2002 with Johansen pleading not-guilty to charges under s 145(2) of the Norwegian Criminal Code, which criminalises unauthorised access to data, an offence that carries a maximum penalty of two years in prison. The defence argued that, since Johansen owned the DVDs he sought to access, and as he was entitled to access the data on them, no illegal access to information was obtained. Further, they argued that it is legal under Norwegian law to make back-up copies of such data for personal use, something which had been rendered impossible by the CSS system. On 7 January 2003, the court gave its verdict. In acquitting Johansen of all charges Judge Irene Sogn ruled that there was 'no evidence' that either Johansen or others had used the DeCSS decryption code illegally. ØKOKRIM filed an appeal. On 28 February the Borgarting Appellate Court agreed to hear the case, and on 5 March they announced that arguments filed by the movie industry and additional evidence presented by them merited another trial. Johansen's second trial began in Oslo on 2 December 2003 with the judgment of the court being issued on 22 December. Again the court noted that Norwegian copyright law allowed for the making of a back-up copy of digital information, such that if the original became damaged the user could recover their data. They noted that any attempt to block such copying could not be legal under Norwegian law: 'A prohibition against copying would limit the right of the consumer compared to the Copyright Act s 12, which permits reproduction of copies of published works for private use when this is not commercial. The appellate court finds that unilateral conditions with respect to a use which is permitted according to the Copyright Act cannot be held valid.'[102] As a result, Jon Johansen was again acquitted. Finally, on 5 January 2004, ØKOKRIM announced that it would not appeal the case any further: the risk of imprisonment was removed from over the head of the young Jon Johansen. The story of DeCSS was though far from over.

The DVD Copy Control Association had also been busy in the US. On 14 January 2000, eight major Hollywood studios[103] raised an action against three individuals and one website operator,[104] claiming the defendants were illegally linking their sites to those that posted the DeCSS code, in breach of the Digital Millennium Copyright Act (DMCA). At this point it is perhaps a good idea to take a short excursion into the origins of the Act itself before examining the outcome of the case. The DMCA was enacted to give effect to Articles 11 and 12 of the WIPO Copyright Treaty (WCT), which had been adopted in Geneva in December 1996. Despite the apparent innocuousness of the Act, commentators such as Pamela Samuelson[105] and Dan Burk and Julie Cohen have noted that 'the DMCA's anti-circumvention measures go well beyond what the treaty requires'.[106] In fact a quick comparison of the two provisions demonstrates the issue. Article 11 of the WCT states:

> Contracting Parties shall provide adequate legal protection and effective legal remedies against the circumvention of effective technological measures that are used by authors in connection with the exercise of their rights under this Treaty or the Berne Convention and that restrict acts, in respect of their works, which are not authorized by the authors concerned or permitted by law.

Article 12 provides that:

> Contracting Parties shall provide adequate and effective legal remedies against any person knowingly performing any of the following acts knowing, or with respect to civil remedies having reasonable grounds to know, that it will induce, enable, facilitate or conceal an infringement of any right covered by this Treaty or the Berne Convention:

---

102 Decision of the Borgarting Appellate Court, 22 December 2003. English version available at: *www.efn.no/DVD-dom-20031222-en.html.*
103 Universal Studios, Paramount Pictures, Metro-Goldwyn-Mayer, Tristar Pictures, Columbia Pictures, Time-Warner, Disney and 20th Century Fox.
104 Shawn Reimerdes, Eric Corley, Roman Kazan and 2600 Enterprises Inc.
105 Samuelson, P, 'Intellectual Property and the Digital Economy: Why the Anti-Circumvention Regulations Need to Be Revised', 1999, 14 *Berkeley Technology Law Journal* 519.
106 Burk and Cohen, see above, fn 63, p 65.

(i) to remove or alter any electronic rights management information without authority;

(ii) to distribute, import for distribution, broadcast or communicate to the public, without authority, works or copies of works knowing that electronic rights management information has been removed or altered without authority.

When we compare this with the text of §1201(a)(2) of the Copyright Act, as inserted by the DMCA, the extent of the difference becomes clear:

'No person shall . . . offer to the public, provide or otherwise traffic in any technology . . . that:

(A) is primarily designed or produced for the purpose of circumventing a technological measure that effectively controls access to a work protected under [the Copyright Act];

(B) has only limited commercially significant purpose or use other than to circumvent a technological measure that effectively controls access to a work protected under [the Copyright Act]; or

(C) is marketed by that person or another acting in concert with that person with that person's knowledge for use in circumventing a technological measure that effectively controls access to a work protected under [the Copyright Act].'

Whereas Articles 11 and 12 of the WCT require protection of the copyright in the underlying works by protecting the digital locks attached to the content and by prohibiting the trade in content, which has been unlawfully decrypted, the DMCA goes much further by attacking the decryption keys themselves. Logically, under the WCT it should be unlawful to use the DeCSS key where the user does not have permission or is not otherwise authorised to do so, such as by fair dealing provisions, or to distribute decrypted content without permission or authority. The WCT does not suggest that merely distributing the DeCSS code should be outlawed, yet this is exactly what the DMCA does. This distinction between protecting the copyright in the underlying material and protecting the locks themselves is at the heart of an extensive debate in the US,[107] was instrumental in the formation of the Creative Commons

---

107 Eg Burk and Cohen, see above, fn 63; Samuelson, see above, fn 105; Samuelson, P, 'DRM {and, or, vs.} the Law', 2003, 46 *Communications of the ACM*, 41; Benkler, Y, 'The Battle over the Institutional Ecosystem in the Digital Environment', 2001, 44 *Communications of the ACM*, 84; Therien, J, 'Exorcising the Specter of a "Pay per Use" Society: Toward Preserving Fair Use and the Public Domain in the Digital Age', 2001, 16 *Berkeley Technology Law Journal*, 979, available at: *http://www.law.berkeley.edu/journals/btlj/articles/vol16/therien/therien.pdf*.

movement and led to Eric Eldred and Lawrence Lessig challenging the Sonny Bono Copyright Term Extension Act, which was passed the same week as the DMCA, as they feared it was a move, when take alongside the DMCA, toward perpetual and complete control of creative works.[108] The extensive provisions of the DMCA were also to prove fatal to the hopes of the four defendants in the DeCSS case. In his opinion, District Judge Lewis Kaplan noted that the: 'defendants clearly violated Section 1201(a)(2)(A) by posting DeCSS to their website'[109] and that their defence that DeCSS 'was written to further the development of a DVD player that would run under the Linux operating system, as there allegedly were no Linux compatible players on the market at the time' was rejected as:

> [The] inescapable facts are that (1) CSS is a technological means that effectively controls access to plaintiffs' copyrighted works, (2) the one and only function of DeCSS is to circumvent CSS, and (3) defendants offered and provided DeCSS by posting it on their website. Whether defendants did so in order to infringe, or to permit or encourage others to infringe, copyrighted works in violation of other provisions of the Copyright Act simply does not matter for purposes of Section 1201(a)(2).

> The offering or provision of the program is the prohibited conduct – and it is prohibited irrespective of why the program was written, except to whatever extent motive may be germane to determining whether their conduct falls within one of the statutory exceptions.[110]

Two of the defendants, Eric Corely and 2600 Enterprises Inc., appealed the case to the Court of Appeals for the Second Circuit. They claimed that 'subsection 1201(c)(1) of the Copyright Act, which provides that "nothing in this section shall affect rights, remedies, limitations or defenses to copyright infringement, including fair use, under this title," should be read to allow the circumvention of encryption technology protecting copyrighted material when the material will be put to "fair uses" exempt from copyright liability'.[111] The Court rejected this claim, simply stating: 'we disagree that subsection 1201(c)(1) permits such a reading. Instead, it clearly and simply clarifies that the DMCA targets the *circumvention* of digital walls guarding copyrighted material (and trafficking in circumvention tools), but does not concern itself with the *use* of those materials after circumvention has occurred. Subsection 1201(c)(1) ensures that the DMCA is not read to prohibit the "fair use" of information just because that information was obtained in a manner made

---

108   *Eldred v Ashcroft* 537 US 186 (2003).
109   *Universal Studios v Reimerdes* 111 F Supp 2d 294 (SDNY 2000), p 319.   110   Ibid.
111   *Universal City Studios Inc v Corley* 273 F 3d 429, p 443.

illegal by the DMCA.'[112] The Court then went on to be equally dismissive of the claim that the First Amendment allowed the Appellants to post the DeCSS Code and links to sites that hosted the code, noting that: 'a content-neutral regulation need not employ the least restrictive means of accomplishing the governmental objective. It need only avoid burdening "substantially more speech than is necessary to further the government's legitimate interests". The prohibition on the Defendants' posting of DeCSS satisfies that standard.'[113] Thus in the view of the Court, Congress had correctly balanced the First Amendment interests of code such as DeCSS, with the need to protect copyright holders to incentivise the further creation of creative works. It is important to note the Court does not deny the existence of First Amendment protection, it merely points out that such protection cannot be allowed to ride roughshod over the values of others. It is perhaps this part of the decision that convinced the appellants that it would not be worthwhile to appeal to the Supreme Court: the media distribution industries had won a vital victory in the battle against DRM crackers.

The decision confirmed that the DMCA could be used by copyright holders to legally control not only the distribution of copyright protected content; they could also control the copying and distribution of decryption keys, whatever the reason was for the creation of the key in the first place. Emboldened by this decision, content suppliers have sought to extend the scope of this principle in a number of ways. Several more cases have been brought in the US under the DMCA, including the highly publicised case of *United States v Elcomsoft*,[114] which ended in an unexpected setback for the copyright holders when a Federal Jury in San Francisco found Russian programmer, Dmitry Sklyarov, and his employer, Elcomsoft, not guilty of criminal charges under the DMCA. Sklyarov and Elcomsoft had developed a product known as the Advanced eBook Processor (AEBPR), a Windows-based program that allows a user to remove use restrictions from Adobe Acrobat PDF files and files formatted for the Adobe eBook Reader. At an earlier hearing District Judge Ronald Whyte had dismissed an application for dismissal from Elcomsoft, noting that: 'Protecting the exclusive rights granted to copyright owners against unlawful piracy by preventing trafficking in tools that would enable widespread piracy and unlawful infringement is consistent with the purpose of the Intellectual Property Clause's grant to Congress of the power to "promote the useful arts and sciences" by granting exclusive rights to authors in their writings.'[115] The later jury decision was therefore quite unexpected and was hailed by the Electronic Frontier Foundation (EFF) as 'a strong message to federal prosecutors who believe that tool makers should be thrown in jail just because a copyright owner doesn't like

---

112   *Universal City Studios v Corley*, above, fn 111 (emphasis in the original).
113   Ibid, p 455.
114   203 F Supp 2d 1111 (ND Calif 2002).   115   Ibid, p 1138.

the tools they build'.[116] This setback aside though, copyright holders have continued to prosecute claims under the DMCA,[117] and have continued to press both in the US and elsewhere for strong anti-circumvention regulations. Their lobbying has met with almost universal success. In the EU the Directive on Copyright and Related rights in the Information Society[118] restricts all acts of circumvention,[119] bans the importation, sale, rental or possession for commercial purposes of all tools designed to allow circumvention of encryption systems,[120] and the distribution of content from which a rights management system has been removed.[121] Meanwhile, on 18 May 2004, Australia and the US signed a bilateral free-trade agreement.[122] Article 17.4.7 of this agreement requires both signatories to:

> Impose civil and criminal penalties, regardless of whether copyright in the materials protected by the TPM is actually breached, on any person who:
>
> (1) knowingly, or having reasonable grounds to know, circumvents without authority any effective technological measure that controls access to a protected work, performance, or phonogram, or other subject matter, or
> (2) manufactures, imports, distributes, offers to the public, provides, or otherwise traffics in devices, products, or components, or offers to the public, or provides services that:
> (a) are promoted, advertised, or marketed for the purpose of circumvention of any effective technological measure,
> (b) have only a limited commercially significant purpose or use other than to circumvent any effective technological measure, or
> (c) are primarily designed, produced, or performed for the purpose of enabling or facilitating the circumvention of any effective technological measure.

The US is already compliant thanks to the DMCA, but the Australian Copyright Act 1968 currently only prohibits, among other things, the importation and manufacturing of circumvention devices; it does not prohibit actual circumvention. To ensure compliance with the free-trade agreement the Australian Government is currently amending the Copyright Act,

---

116 EFF Press Release 17 December 2002, available at: *www.eff.org/IP/DMCA/US_v_Elcom soft/20021217_eff_pr.html*.

117 Eg *Lexmark v Static Controls* 253 F Supp 2d 943 (ED Ky 2003); *Chamberlain Group Inc v Skylink Technologies Inc* (Federal Circuit, 31 August 2004), available at: *http://laws.lp.find law.com/fed/041118.html*.

118 Dir. 2001/29/EC.   119 Art 6(1).   120 Art 6(2).   121 Art 7(1).

122 Full text of the Agreement may be found at: *www.ustr.gov/Trade_Agreements/Bilateral/ Australia_FTA/Final_Text/Section_Index.html*.

thus anti-circumvention provisions are now, or are about to be, found in the laws of the US, the EU, Australia and many other countries worldwide including Bahrain,[123] Singapore[124] and Morocco.[125]

The market position of content providers is starting to look unassailable. Not only have courts in the US, Australia and elsewhere taken action to close down third-party intermediaries such as Napster and Grokster, but legislative developments and litigation have succeeded, at least to some degree, in protecting the code locks that protect content. Surely with such unprecedented levels of protection the media distribution industries would be happy to accept that although a small amount of leakage may occur, the market in digital creative products was secure? Unfortunately, despite nearly 10 years of constant lobbying and litigation the media entertainment industries have had little effect on the black market in digital entertainment files. No matter how many P2P systems are shut down or brought in house, the problem remains as the technical know-how to build a P2P network is widely available and there will be always be others willing to take their place. Further, the DMCA and allied legislative provisions around the world have had little effect on crackers who continue to reverse engineer and distribute both decryption keys and 'cracked' content. A quick Google search reveals over 1,000 websites currently offering DeCSS downloads, while cracks for everything from Apple's Fair Play system to Windows Media Player's DRM 10 and Adobe DRM are widely available to download from both websites and through P2P and IRC systems. Thus despite a sustained legal attack on file-sharers, crackers and P2P service providers in the period 1996–2005, statistics reveal that the number of file-sharers using the most popular P2P services actually increased by nearly 85 per cent from a little over 4.6 million users in January 2003 to 8.5 million users in April 2005.[126] It seems that nothing the lawyers, legislators and code-designers do will slow down the growing black market in illegal file-sharing. In apparent desperation the industry has resorted to suing individual file-sharers. In so doing it is taking a huge commercial risk, as although there is little doubt these individuals are acting unlawfully, and that their actions are detrimentally effecting the market in creative consumer goods, the risks of pursuing its own customers, many of whom are students

---

123 Imposed by Art.14.4.7 of the US-Bahrain FTA of 14 September 2004. Text available at: *www.ustr.gov/assets/Trade_Agreements/Bilateral/Bahrain_FTA/final_texts/ asset_upload_file211_6293.pdf.*

124 Imposed by Art.16.4.7 of the US-Singapore FTA of May 6 2003. Text available at: *www.ustr.gov/assets/Trade_Agreements/Bilateral/Singapore_FTA/Final_Texts/ asset_upload_file708_4036.pdf.*

125 Imposed by Art.15.5.7 of the US-Morocco FTA of 15 June 2004. Text available at*www.ustr.gov/assets/Trade_Agreements/Bilateral/Morocco_FTA/FInal_Text/asset_upload_ file797_3849.pdf.*

126 *www.slyck.com/news.php?story=763.* Since then this figure has remained reasonably static at around 8.5 million users.

and young adults, and who can therefore be expected to be customers for a considerable period, should not be underestimated. Figures released by the International Federation of the Phonographic Industry (IFPI) in April 2006 reveal at that time that the IFPI and its member institutions had raised 24,176 actions against individual file-sharers in 17 countries.[127] They claim the campaign to be a success noting that 'three million people have either reduced or stopped illegal file-sharing in Europe' and that 'legal buying is more popular than P2P in Europe's two major digital markets, Germany and the UK'.[128] Yet despite these claims of success there is a concomitant degree of bad publicity including the reported case of a 12-year-old girl, who was on the end of a Recording Industry Association of America (RIAA) writ.[129] The truth is that for every file-sharer who receives a writ from the IFPI or RIAA, deletes the infringing material from their hard drive and pays a settlement, there is another to take their place. While it is true that the absolute numbers of files being downloaded are falling, and that the subscribers to legal music distribution sites are rising, to claim this is purely as a result of the RIAA/IFPI action obfuscates the true picture.

The market functions – but only so far!

It is not legal successes driving these changes – it is the market; in particular, the runaway success of the iTunes music store, which provides efficient and convenient downloads for the 50 million iPod users worldwide.[130] With greater availability of affordable and legal online music stores such as iTunes or the relaunched Napster service, the casual infringer who used P2P clients such as KaZaA or StreamCast to build his/her digital music library simply because it was more convenient than buying a CD, then downloading it to their MP3 player, is switching to legal download services. It is this migration of the 'casual' downloader, which helps explain both the drop in illegal downloads and the associated rise in legal downloads. These figures clearly demonstrate that if the media entertainment industries offer customers a convenient, and competitive, alternative to illegal file-sharing sites, a substantial proportion of customers will make the switch. This suggests that instead of pursuing almost constant litigation against all parties to the illegal trade in copied audio or video files, the money used to fund this litigation should be invested directly in further development of the legal download market. However, it has to be acknowledged that no matter what approach the

127 www.ifpi.org/site-content/press/20060404a.html.
128 www.ifpi.org/site-content/press/20060404c.html.
129 http://civilliberty.about.com/cs/onlineprivacy/a/blDMCA091103.htm.
130 By December 2005 Apple had sold 42 million iPods (see http://reviews.cnet.com/4531–10921_7–6416165.html). Recent Q1, 2006 figures reveal 8 million sales in Q1 2006 making a total of 50 million iPods sold. See www.apple.com/pr/library/2006/apr/19results.html.

industry takes, there will remain a hard-core of file-sharers who will not be tempted by the offer of convenient, secure legal downloads and who will ignore the threat of legal action. Thus it would be naive to assume that illegal file-sharing will ever be completely eradicated. Once one accepts, though, that piracy is the constant companion of creative products, then a more balanced approach to its control may be taken. The media distribution industries need first, though, to accept that the underlying economics of the market for their products has changed with the disintermediation of content and carrier. At present, there appears to be little likelihood of this occurring. Currently, what we see, when we look at how the industry has responded to the challenge of digitisation, is a series of actions that mirror the standard response pattern of established industries to the challenge of disruptive technologies, which can be traced as far back as the printing press itself.

The first stage is to attempt to curtail or seek control of the technology. For example when the Gutenberg press proved to be successful the Papal Court contemplated licensing the production of printing presses in an attempt to control their spread. Similarly, the first response of the fledgling rail industry to the nascent challenge of the motor vehicle was to press Parliament for a series of Acts, to control and restrain the new technology.[131] The second stage is to seek to implement legal controls to protect the status quo. This can clearly be seen in the actions of the film industry in dealing with the threat of the video cassete recorder (VCR). The industry, through Universal Studios and Walt Disney studios, initially raised an action in the District Court for the Central District of California claiming that as Sony's VCR device could be used for copyright infringement, they were thus liable for any infringement that was committed by its purchasers. Following a defeat for the industry on appeal before the Ninth Circuit, they changed their line of attack slightly by seeking Congressional support for their campaign. This led to the famous testimony of Jack Valenti, President of the Motion Picture Association of America (MPAA), before the House of Representatives hearing on Home Recording of Copyrighted Works, where he stated that 'the VCR is to the American film producer and the American public as the Boston strangler is to the woman home alone'.[132] In so doing he hoped not only to obtain Congressional support to overturn the decision of the Court of Appeals for the Ninth Circuit in the case, but also to gain public support for the campaign

---

131 The Acts were (in order): The Locomotive Act 1861, which limited the weight of steam driven vehicles to 12 tonnes and imposed a speed limit of 10 mph; The Locomotive Act 1865, which set a speed limit of 4 mph in the country and 2 mph in towns; The Locomotive Act 1865, which provided for the famous 'man with a red flag' who had to walk 60 yards ahead of each vehicle. This enforced a walking pace on the driver of the motor vehicle; and The Locomotive Amendment Act 1878, which made the red flag optional subject to local regulations. Collectively these Acts are sometimes called 'The Red Flag Acts'.

132 12 April 1982. Available at: *http://cryptome.org/hrcw-hear.htm*.

of the MPAA. In the event, his testimony was disastrous for the MPAA campaign, with members of the public outraged at his clumsy metaphor and with Congress eventually refusing to be drawn into the legal dispute. The case, as we know, went on to be heard by the Supreme Court where the film industry was famously defeated.[133]

A similar approach may be seen in the current approach of the music industry to file-sharing technologies. They too have taken legal action, against Napster, against KaZaA and against Grokster/Streamcast to seek to maintain their position in the market for music and sound recordings. Also, like the MPAA in 1982, they seek Congressional support to protect the status quo. Not only have we had the DMCA, but we have also seen proposals for some quite outlandish Congressional Bills, including the draft Inducing Infringement of Copyrights Act 2004, introduced by Senator Orrin Hatch and which, if enacted, would result in the creation of a new offence of inducement to infringe where the defendant 'intentionally aids, abets, induces, or procures the infringement of others' – a measure aimed directly at P2P service providers, and the draft Protecting Intellectual Rights Against Theft and Expropriation Act of 2004, or PIRATE Act, which would allow the Justice Department to pursue a civil case against file-sharers.[134] Although the music industry has had a much greater degree of success, both before the courts and Congress, than the film industry had in 1982, the challenge posed by the technology is also much greater, meaning that these successes aside, it is, as we have seen, unlikely that the legal system can support the status quo in the market for music for much longer. This means that the music industry (and soon the film and TV industries) must plan to move quickly to the third stage, which is to adapt to the distruptive technology and accommodate it into their business models. This stage can provide exciting new opportunities and income streams as the film industry was to find when it started exploring the possibilities of home VCR technology; far from being the Boston Strangler of the industry, it turned out to be its Fairy Godmother with the home rental and sell-through video markets not only providing valuable new markets to develop, but also signalling a renaissance at the box office. The same opportunity awaits the media distribution industries in the digital marketplace. The market will eventually settle on a fair price for digital media files, but only once the market is allowed to mature. Continued legal intervention in the market will only serve to further undermine the market, its values and its dynamics. If allowed to develop, the market will provide exciting new opportunities for suppliers and consumers alike, but as with all other markets, attempts to interfere in the market may lead to unexpected, and unpredictable, results.

133 *Sony v Universal Studios*, above, fn 74.
134 It should be noted that if enacted, the PIRATE Act would shift the economic burden of protecting private property rights from the rights-holder to the government, thus public funds would be used to protect private property interests.

Chapter 7

Cyber laws and cyber law-making

The Internet interprets the US Congress as system damage and routes around it.

Jeanne DeVoto

In his famous address, *A Declaration of the Independence of Cyberspace*,[1] arch-cyber-libertarian John Perry Barlow informed the 'Governments of the Industrial World' that 'you [do not] possess any methods of enforcement we have true reason to fear . . . our identities have no bodies, so, unlike you, we cannot obtain order by physical coercion'. This powerful challenge to traditional regulators reflected the cyber-libertarian ethos, discussed in Chapter 1, that laws could not succeed in cyberspace due to the physical nature of the place. As we saw in that earlier discussion, the cyber-libertarian ethos was founded upon the twin foundations of the 'unregulability of bits'[2] and the ability of internet users to transcend borders without challenge.[3] Laws, like all other regulatory control systems, consist of three elements: a director, a detector and an effector.[4] Directors are standards, detectors are means of detecting some deviation from the standards and effectors are mechanisms for pushing deviant behaviour back towards the standard. The 'unregulability of bits' thesis suggested to cyber-libertarians such as Barlow that there were no functional effectors that could be used to enforce traditional laws in cyberspace as effectors used in traditional legal-regulatory control systems such as

---

1 Made at Davos, Switzerland, 8 February 1996, available at: *www.eff.org/~barlow/Declaration-Final.html*.

2 The unregulability of bits thesis, discussed above at pp 6–7 was based on the hypothesis that computer data as mere flows of digital information, which had no physicality or mass, could not be effected by traditional models of regulation which are based upon physicality.

3 Johnson, D and Post, D, 'Law and Borders – The Rise of Law in Cyberspace', 1996 48 Stan LR 1367.

4 Murray, A and Scott, C, 'Controlling the New Media: Hybrid Responses to New Forms of Power' 2002 65 MLR 491, p 502.

imprisonment, fines and orders for specific performance would prove to be ineffective against digital personae who have no physicality, identity or fiscal funds. Thus traditional laws and law-making, based upon the concept physicality, were rendered ineffective in cyberspace in the true sense of the word. In addition, the lack of borders within cyberspace suggested to cyber-libertarians that traditional state sovereignty, based upon notions of physical borders, would be undermined in cyberspace as individuals could move seamlessly between zones governed by differing regulatory regimes in accordance with their personal preferences. This, the cyber-libertarians believed, would foster regulatory arbitrage and undermine traditional hierarchically structured systems of control. Basing their beliefs upon these two foundational concepts the cyber-libertarians argued that the lack of a physical presence to cyberspace effectively undermined traditional laws and law-making, which are based upon the concept of physicality of goods and persons.[5]

As we know, the cyber-libertarian thesis was quickly undermined by the approach taken by the cyberpaternalists who demonstrated the propensity for the development of private regulatory regimes[6] within 'sovereign' cyberspace. The cyberpaternalist thesis was demonstrated by Joel Reidenberg who identified two types of private regulatory systems: (1) regimes based upon contractual agreements such as those between ISPs and customers; and (2) regimes built upon the network architecture such as the technical standards promulgated by bodies like the Internet Engineering Task Force (IETF). Reidenberg demonstrated how these systems could, through the application of design controls, act as proxies for courts and law-enforcement authorities.[7] Reidenberg's concept of control through technology, which he titled the *Lex Informatica*, went on to be developed and applied to a variety of regulatory problems within cyberspace by a number of eminent computer scientists and cyberlawyers, including Mark Stefik,[8] Mark Lemley,[9] Cass Sunstein[10] and most famously, Lawrence Lessig.[11] The fact that the cyberpaternalist school of thought has become the dominant school of cyber-regulatory theory over the last 10 years should not though obscure the relevance of the

5 As Barlow claimed, 'Your legal concepts of property, expression, identity, movement, and context do not apply to us. They are all based on matter, and there is no matter here.' See *A Declaration of the Independence of Cyberspace*, see above, fn 1.
6 By private regulatory regimes I mean those systems of regulation or control promulgated by private enterprises with or without the support or permission of the state or the democratic agreement of the citizens of the state.
7 See above, pp 8–9.   8 Stefik, M, *The Internet Edge*, 1999, Cambridge, MA: MIT Press.
9 Lemley, M, 'Place and *Cyberspace'*, *2003, 91 Calif L Rev* 521.
10 Sunstein, C, *Republic.com*, 2001, Princeton, NJ: Princeton UP.
11 Lessig, L, *Code and Other Laws of Cyberspace*, 1999, New York: Basic Books; Lessig, L, *The Future of Ideas: The Fate of the Commons in a Connected World*, 2001, New York: Random House; Lessig, L, *Free Culture: How Big Media Uses Technology and the Law to Lock Down Culture and Control Creativity*, 2004, New York: Penguin.

cyber-libertarian argument. In a sense, the cyber-libertarians lost the debate due to a technicality. They assumed that a lack of effectiveness of traditional legal-regulatory control systems meant freedom within the environment of cyberspace: the cyberpaternalists demonstrated that in fact it simply facilitated the substitution of alternative private regulatory systems. But the underlying arguments of the cyber-libertarians remain: it is extremely difficult and costly to enforce traditional legal-regulatory control systems within cyberspace, due to a variety of factors including a relative degree of anonymity, lack of physicality, digitisation of content, environmental plasticity[12] and the international or cross-border nature of the network. Therefore, although the cyber-libertarians did not succeed in demonstrating an inherent unregulability in the network, they did demonstrate the problems lawmakers would face, and continue to face, in applying traditional legal-regulatory control systems in the network environment. This chapter will examine these difficulties and some of the attempted solutions to these problems by examining the effectiveness of law as a direct regulator within cyberspace. The following chapter will then examine the role law plays as part of a hybrid regulatory system.

Law, indecency, obscenity, speech

As was discussed in Chapter 5, one of the greatest challenges regulators face in cyberspace is the management and control of pornographic content.[13] Pornography comes in many shapes and forms, and often there is no clear legal definition as to when material is obscene, even within individual states. For instance, the legal definition of obscenity in England and Wales dates from 1959 and is extremely open-ended. By s 1(1) of the Obscene Publications Act (OPA) 1959 an article is deemed to be obscene if:

> . . . its effect or (where the article comprises two or more distinct items) the effect of any one of its items is, if taken as a whole, such as to tend to deprave and corrupt persons who are likely, having regard to all relevant circumstances, to read, see or hear the matter contained or embodied in it.

As can be seen, this definition gives a great deal of discretion to the courts. What exactly is likely to 'deprave and corrupt' persons? Is it the sight of unclothed flesh? Is page 3 of *The Sun* likely to deprave and corrupt?[14] How

---

12 See above, pp 35–43.   13 See above, pp 157–63.

14 For overseas readers this question may need contextualisation. *The Sun* is a national daily newspaper in the UK owned by Rupert Murdoch's News Corp. As part of its rebranding in the early 1970s it placed a picture of a topless model on page 3 of the paper on 17 November 1970 and has been doing so on a regular basis ever since.

about the content of a top-shelf magazine? Does there need to be some kind of sexual act? If so, of what level of explicitness? These are vitally important questions to be answered as they help us to define the border between merely offensive material, which is legal in the UK, and obscene material which, under s 2 of the OPA 1959, it is an offence to publish or to possess for publication for gain.

Perhaps due to the age of the law, or perhaps in what is a sign that the UK is a diverse and multicultural society, there is a disparity between what is deemed to be obscene under the OPA, and what may be deemed to be offensive to a significant proportion of the general public. As a result of this disparity there have been extensive campaigns to amend and update the UK's laws on pornography by groups such as the Campaign Against Pornography,[15] the National Viewers and Listeners Association,[16] and the National Campaign for the Reform of the Obscene Publications Acts.[17] For

15 The Campaign Against Pornography was established in 1987 as a response to MP Clare Short's Bill to ban erotica such as 'Page 3' in national newspapers. Its aims were 1) to promote equal opportunities for women and oppose discrimination; 2) to preserve and protect the health and safety of women and children through working towards the elimination of violent crimes against them; 3) to undertake research into the links between pornography and violence against women and children and their position in society and to publish the results of such research; 4) to gather information related to the production, publication, distribution and consumption of pornography and its effects on women and children's lives and their position in society; 5) to provide facilities for the relief of women and children affected by pornography; 6) to raise funds and receive contributions; 7) to co-operate and collaborate with any voluntary organisation and statutory authorities having similar objects and to exchange information and advice with such bodies. The group was incorporated in April 1989 and undertook a series of educational programmes that included training events, school workshops and seminars. It focused on a number of specialised campaigns in addition to its general aims with such activities as helping prevent the launch of a satellite channel in 1993, protesting over a number of advertisements in the press and picketing local events. It also undertook a letter writing campaign in 1995, by which time it had a number of local groups affiliated to it throughout the country. However, it ended its activities in the late 1990s through the effects of a financial crisis.

16 The National Viewers and Listeners Association, now mediawatch-uk, was set up by Mary Whitehouse to pressurise broadcasting authorities to improve their public accountability and to explain their policies on standards of taste and decency. As part of its campaign profile Mrs Whitehouse consistently campaigned for effective amendment to the Obscene Publications Act 1959. However (as noted by mediawatch-uk on their website), 'because of political and judicial indifference, this objective remains to be realised but mediawatch-uk continues, along with others, to press the Government at every opportunity on this important issue'.

17 The National Campaign for the Reform of the Obscene Publications Acts was formed in April 1976 by David Webb and others who shared both his vehement opposition to censorship and also his concern that a totally unrepresentative minority of the general public was, or at least appeared to be, becoming increasingly successful in its efforts to force its beliefs and moral standards on others. The NCROPA believes that this is totally alien to the concept of what people in this country have come to expect of a supposedly free society and presents an intolerable curtailment of individual liberty and the freedom of expression. Furthermore,

their part, the courts have endeavoured not to get involved in this debate and will refer any attempt to use case law to update or reinterpret the law back to the wording of the Act. For instance in *R v Anderson & Others*,[18] the case of the infamous *Oz Magazine* school kids issue,[19] the Court of Appeal upheld an appeal against conviction under the Act and found that the trial judge had misdirected the jury in suggesting that 'obscene' included in its meaning the dictionary sense of repulsive, filthy, loathsome or lewd. The Lord Chief Justice, Lord Widgery, gave the opinion of the Court and stated clearly for the record:

> There is no doubt in our judgment but that obscene in its context as an alternative to indecent has its ordinary or as it is sometimes called dictionary meaning. It includes things which are shocking and lewd and indecent and so on. On the other hand, in the Obscene Publications Act 1959, there is a specific test of obscenity, and in charges under that Act it is this test and this test alone which is to be applied, and it is in [the] form [of] Section 1 of the Act.[20]

Similarly, in the case of *R v Uxbridge Justices ex parte David Webb*,[21] the Divisional Court refused an application for an order of mandamus, which the applicant, David Webb, sought in order to compel the justices of Uxbridge Magistrates' Court to state a case for appellate consideration in relation to an earlier order they had made for the forfeiture of six video cassettes, showing explicit homosexual activity, which he had been given in the Netherlands and which he intended to view privately in connection with his work for the National Campaign for the Reform of the Obscene Publications Acts. Mr Webb, as one might expect, raised several issues before the Court, including questioning 'whether the interpretation of the word "obscene" advanced by the Customs and Excise during the course of the argument at the Magistrates' Court and adopted by that court, was correct'[22] and claiming 'that the Magistrates' Court ignored the provisions of Article 10(1) and (2) of the European Convention of Human Rights'.[23] Throughout, the Court treated Mr Webb's claims with respect, but were steadfast in their refusal to be drawn into what seemed ultimately to be a policy argument. In response to Mr Webb's first claim, Glidewell LJ was clear:

---

it is in direct contravention of both the United Nations Universal Declaration of Human Rights and the European Convention on Human Rights [excerpted from the NCROPA website at: *http://freespace.virgin.net/old.whig/NCROPA/whatis.htm*].

18 [1972] 1 QB 304.

19 For more on the case see Carlin, G and Jones, M, *The Rupert Bear Controversy: Defence and Reactions to the Cartoon in the OZ Obscenity Trial*, available at: *http://pers-www.wlv.ac.uk/~fa1871/rupage.html*

20 Ibid, pp 311–2.   21 [1994] 2 CMLR 288.   22 Ibid, p 291.   23 Ibid.

There is no definition of the word 'obscene' in either the 1876 or 1979 Customs Act. However, the decision of the House of Lords in R. v Henn [1981] A.C. 850 is clear authority for the proposition that in interpreting that word in the Customs Acts, the definition of the word contained in section 1 of the Obscene Publications Act 1959 is to be adopted. The Magistrates adopted that definition. As I have said there is clear authority for the proposition that they were right to do so and there is no issue of law on which the Court could conceivably find in Mr. Webb's favour that that was a wrong definition because this court was bound by the House of Lords' decision in Henn.[24]

In response to his second claim he was equally steadfast:

Article 10(1) provides in general terms for freedom of expression and publication. But ... Article 10(2) of the Convention of Human Rights contains an exception to or restriction on the width of Article 10(1), because it provides that the law of a Member State may include provisions which are 'necessary in a democratic society for the protection of morals'. There is no doubt that it can properly be held that the provisions of domestic legislation whether one agrees with them or not, whether one thinks that domestic legislation of the United Kingdom ought to be more akin to that of Germany, Denmark or Holland, is not the point. It does not contravene the European Convention of Human Rights in so far as the exception in Article 10(2) expressly provides that there may be such legislation.[25]

What we see, both from the definition of obscenity given in s 1 of the OPA, and from the decisions of the courts in *R v Anderson* and *ex parte Webb* is a desire on the part of Parliament and the judiciary to remain apart from the public debate on questions of morality, decency and freedom of expression. This approach it should be said is most sensible because questions of indecency and obscenity appear to divide public opinion like no other, and no matter which approach the law took it would be highly criticised by those on one side of the debate. By continuing to apply, without amendment, the dated yet flexible definition of obscenity found in the 1959 Act, the law finds itself able to adapt to changes in society. What was likely to deprave and corrupt in 1959 is unlikely to deprave and corrupt in modern society. It is for instance unimaginable today that the Director of Public Prosecutions (DPP) would take action against anyone publishing copies of *Lady Chatterley's Lover*,[26] or that in 1961 the British Board of Film Censors (the predecessors to the current British Board of Film Classification) would allow for the Michael

24 See fn 18, above.   25 Ibid, p 292.   26 *R v Penguin Books Ltd* [1961] Crim LR 176.

Winterbottom film *Nine Songs*, which features unsimulated sex onscreen to have been passed uncut for theatrical release. This is not to suggest, or even pretend that the law in the UK is ideal. In fact, it may be argued that by doing so much 'fence-sitting', the lawmakers and the courts leave themselves open to criticism from both sides. Those who find the current settlement to be too permissive argue that the definition of obscenity is too lax and is in dire need of being tightened up, lest British society debase itself further. Those who find the current settlement too draconian, argue that the British obscenity laws are among the least permissive in western Europe and that they interfere unnecessarily with freedom of expression. What is clear is that however the UK legal system deals with the availability and supply of pornographic material, a large portion of the UK population will be alienated. It is this policy tightrope that has held back development of a new Obscene Publications Act.

Indecency, obscenity and cyberspace

The difficulties that lawmakers have faced in the UK are instructive of the kinds of problems that we may face in the online environment. The UK is, as we have seen, a distinct, although large, macro-community.[27] When you expand the extent of the community to be regulated beyond the boundaries of the traditional sovereign jurisdiction, the problem is magnified. In a multinational, multicultural environment such as the internet, cultural, moral and legal variations around the world make it almost impossible to define 'pornographic content' in a global society. As was discussed in Chapter 5, the concept of a global community in cyberspace is sometimes overplayed.[28] Cyberspace is made up of a multitude of micro-communities, reflecting the needs or interests of its members, and these communities have a variety of individual definitions of obscenity. Thus content deemed obscene in a family community such as *myparentime.com* may be deemed to be acceptable in an adult community such as *thephotoforum.com*. Equally, content deemed obscene to a religious community such as *praise.café* would probably not raise an eyebrow in a movie discussion forum such as *outermost.net*. Perhaps most worryingly for parents and for governments are the niche communities which discuss and share hardcore pornography depicting among other things sexual violence (including rape), bestiality and paedophilia. Although such content is almost certain to be classified as obscene by the current laws of any real world State, and by the vast majority of internet users, for the small minority who make up the membership of these communities there is no

27 See above, pp 130–1.
28 See above, pp 141–2. See also Walmsley, D, 'Community, Place and Cyberspace', 2000, 31 *Australian Geographer* 5.

social condemnation of their values while they are within the body of that community. The problem for lawmakers, though, is that these individuals do not reside wholly within that online community: they are also functioning members of real world communities, including the macro-community of citizens of the State in which they live. Lawmakers must deal with the effects of this duality: the actions of these individuals while in the online community can have effect in their real world communities. For example members of such online communities will often create original content to be shared with other members of the community: this usually involves using models and holding photo sessions. Because of the discrepancy in values between the online and real world communities, the individual's moral values become blurred: members of such communities may not recognise the effect that their online actions are having on their relationships with other members of their real world community.[29] How then can the law, deal with this problem?

The overwhelming problem that lawmakers face in dealing with this issue is which standard to apply. As we have already seen, getting the UK community to agree a common standard of decency has proven quite impossible over the last 45 years; to imagine that it might be possible to agree a worldwide standard is fanciful. What is considered sexually explicit, but not obscene, in England may well be considered to be obscene in the Republic of Ireland, and almost certainly material considered obscene in the Islamic Republic of Iran or in the Kingdom of Saudi Arabia would not be felt to be noteworthy in England. Similarly, material which would be considered to be obscene in England would probably not be censored in Germany, Spain or Sweden where a more tolerant approach to erotica and pornographic material is taken. What we are seeing in these differences is a spectrum of obscenity which ranges from extremely conservative to extremely liberal, and upon which individual States position themselves. In general this system has functioned quite effectively in the real world, due to the existence of physical borders and border controls. The easiest way for a State to apply its legal standard of obscenity within its borders is to prevent the importation of materials that offend the standard of that State, while simultaneously criminalising the production of such materials within the State. In the UK, for example it is an offence to import indecent or obscene prints, paintings, photographs, books, cards, lithographic or other engravings, or any other indecent or obscene articles under s 42 of the Customs Consolidation Act 1876, while s 2 of the OPA 1959 criminalises the publication, or possession with intent to publish, an obscene article. The Customs Consolidation Act

29 Recently Donna Hughes noted that: 'the sexual exploitation of women and children is a global human rights crisis that is being escalated by the use of new technologies.' Hughes, D, 'The Use of New Communications and Information Technologies for Sexual Exploitation of Women and Children', 2002, 13 *Hastings Women's Law Journal* 129.

allows the UK to apply effective border controls, allowing for a differential value of obscenity in different States. It allows HM Revenue & Customs to seize obscene items, and where necessary to prosecute those involved in their importation. This control provision continues to apply despite the UK's membership of the EU, with it being held on several occasions that this power subsists in relation to material deemed obscene under the OPA, despite the effects of Articles 28 and 30 (previously, Articles 30 and 36) of the EC Treaty.[30] But these traditional measures are predicated upon the assumption that the items in question will be fixed in a physical medium, and that they will require physical carriage to enter the State. With the advent of the digital age, both these assumptions have been rendered null. The development of a global informational network has dismantled traditional borders: a point which was so eloquently made by David Post and David Johnson in their seminal paper, *Law And Borders – The Rise of Law in Cyberspace*.[31]

> Cyberspace has no territorially based boundaries, because the cost and speed of message transmission on the Net is almost entirely independent of physical location: Messages can be transmitted from any physical location to any other location without degradation, decay, or substantial delay, and without any physical cues or barriers that might otherwise keep certain geographically remote places and people separate from one another. The Net enables transactions between people who do not know, and in many cases cannot know, the physical location of the other party. Location remains vitally important, but only location within a *virtual* space consisting of the 'addresses' of the machines between which messages and information are routed.[32]

> The Net thus radically subverts a system of rule-making based on borders between physical spaces, at least with respect to the claim that Cyberspace should naturally be governed by territorially defined rules.[33]

Thus the traditional concept of border controls is undermined in the digital environment, presenting HM Revenue & Customs and the police with a herculean task. To take but one simple example: if I were to access and view an obscene image on my home computer, which is hosted on a server based in Sweden, the prosecuting authorities would first have to identify that the item is obscene by applying the definition given in s 1 of the OPA.[34] If this could be

---

30 *Conegate Ltd v HM Customs & Excise* [1987] QB 254 (ECJ); *R v Forbes* [2002] 2 AC 512 (HL).
31 See above, fn 3.   32 Ibid, pp 1370–1.   33 Ibid, p 1370.
34 If they could not establish the item to be in breach of the Act, then I would be entitled to view the item under Art 28 of the EC Treaty. See *Conegate Ltd v HM Customs and Excise; R v Forbes*, see above, fn 30.

established they would next have to prove that either I imported the item in breach of the Customs Consolidation Act, or that I possessed the item with intent to publish in breach of s 2 of the OPA 1959. Neither of these claims would necessarily succeed. The second claim would only succeed in relation to members of communities that trade or share images or files: for individuals who merely access and view pornographic websites there would be no intent to further publish or distribute, and therefore no offence under the OPA. The former claim is one mired in extreme complexity. Whereas identification of an importer was relatively straightforward when dealing with physical goods, it becomes much more complex in relation to digital information. The question is – does the consumer import the image into the UK, or is the image imported into the UK by the supplier who then makes it available to the consumer? The answer may at first seem straightforward: if you download an obscene image from a Swedish website then you should be deemed to be the importer. But what if the website appears to be from the UK? Perhaps the supplier is using a UK-based domain name like www.pornoimages.co.uk,[35] and seems to be implying they are based in the UK. In such circumstances does the end-user exhibit sufficient intent and knowledge to be classed as an importer? These difficulties, allied with the intrinsic difficulty of detecting the importation of specific types of digital content,[36] have led to HM Revenue & Customs and the police to focus their limited resources on narrow areas that produce a sound return on their investment. Thus despite surveys which show that more than a third of UK surfers visit pornographic websites,[37] and that 57 per cent of British 9–19-year olds who go online at least once a week have come into contact with online pornography,[38] there have been no prosecutions in England & Wales under either the Customs Consolidation Act or the OPA 1959 for privately viewing obscene material using an internet connection. Instead, the authorities have focused on key issues such as the storing and distribution of child pornography,[39] and prosecuting those who run pornographic websites from overseas servers, but who are resident in the UK and profit from this activity.[40] With the removal of the physical border

35 At the time of writing no site or registration existed in relation to this address.
36 All digital content when imported into the country is in the form of 1s and 0s. To filter out types of content involves a large-scale investment in filtering and firewall software. This is discussed further below, pp 225–7.
37 'One in three view online porn', BBC News, 22 June 2000, available at: *http://news.bbc.co.uk/1/hi/uk/801972.stm*.
38 Livingstone, S and Bober, M, *UK Children Go Online*, 2005, Department of Media & Communication, LSE, available at: *www.lse.ac.uk/collections/children-go-online/UKCGOfinalReport.pdf*.
39 Eg *R v Barry Philip Halloren* [2004] 2 Cr App R (S) 57; *R v Snelleman* [2001] EWCA Crim 1530 and *R v James* [2000] 2 Cr App R (S) 258.
40 Eg *R v Ross Andrew McKinnon* [2004] 2 Cr App R (S) 46 and *R v Stephane Laurent Perrin* [2002] EWCA Crim 747.

between the UK and the rest of the world, internet users were afforded the opportunity to access and view pornography held overseas in the blink of an eye and with little opportunity for the authorities to intercept the content en route. This caused a huge upsurge in consumption and left the authorities with a difficult decision to make. They could either invest large sums to attempt to enforce the law in the digital environment,[41] or they could de facto deregulate adult obscenity and focus their attention on more pressing problems such as child pornography. The UK authorities, recognising the limits of the law as a regulatory tool in relation to this subject, chose to focus their resources on the most harmful content,[42] but other States did attempt to use legal controls to regulate the availability of indecent and obscene materials in cyberspace with quite unpredictable results.

Indecency, obscenity, cyberspace and the US First Amendment

One of the most interesting case studies is the response of the Government of the United States of America to the growing problem of online pornographic content. The people of the US appear to be involved in a much wider moral debate on the appropriateness of indecent and obscene material than the more limited debate we see in the UK. Whereas UK citizens are willing to accept that free expression does not mean limitless freedom to say or do whatever one wishes, US citizens strongly support their First Amendment right to enjoy freedom of speech, even where that right strays into the potentially destructive areas of pornography and hate speech. The role and effect of the First Amendment on US society, political debate and law-making reflects a distinctively American approach to the regulation of complex social issues such as pornography and indecency, hate speech and violent speech. Whereas in western Europe the traditional legal approach to such issues has been to develop a matrix of laws targeted at alleviating a variety of psychological, sociological and political harms attributed to 'words that

---

41 This could either be achieved by the investment of these funds into additional law enforcement personnel or by using the funds to design a technological solution to the problem such as a national firewall or filtering system which would in effect rebuild the natural border in cyberspace. See discussion below at pp 225–7 and the excellent discussion of this subject found in Deibert, R and Villeneuve, N, 'Firewalls and Power: An Overview of Global State Censorship of the Internet' in Klang, M and Murray, A (eds), *Human Rights in the Digital Age*, 2005, London: Glasshouse Press.

42 Currently in the United Kingdom it is illegal to view or possess images of child pornography under s 160(1) of the Criminal Justice Act 1988, while by s 45 of the Sexual Offences Act 2003 a child is defined as anyone under 18 years of age. Also recently the Home Office and Scottish Executive have carried out a consultation on criminalising the possession of violent sexual imagery, see *www.homeoffice.gov.uk/documents/cons-extreme-porn–300805/cons-extreme-porn–300805?view=Binary*.

wound',[43] the US approach has been to develop 'a "marketplace of ideas" with the presumption being that the state must refrain from regulating the content of speech of any kind and instead rely upon the common sense of the people to discover the truth through unrestricted discussion and debate'.[44] The question of whether it is appropriate to apply the First Amendment to sexual/pornographic content has long vexed US scholars and judges. Some scholars have argued that there can never be a true marketplace of speech in relation to pornographic imagery because there is no real freedom of speech for women in a country in which women are relegated to the particular gender roles that society gives them;[45] others, though, argue that pornographic magazines 'consciously attempt to express a view of social and sexual life'.[46] Thus it may be said that, at very least, pornography does support a strong secondary marketplace of ideas, even if the value of the primary marketplace is controversial. Whatever position one holds on the validity of First Amendment protection for pornographic imagery, the law is quite clear. Material of a sexual nature will be protected by the First Amendment unless that material is determined by the court to be obscene.[47] The current test of obscenity was set out in the landmark case of *Miller v California*.[48] In this case the Supreme Court established a three-part test: to be obscene, a judge and/or a jury must determine:

(1) That the average person, applying contemporary community standards, would find that the work, taken as a whole, appeals to the prurient interest; AND
(2) That the work depicts or describes in a patently offensive way, as measured by contemporary community standards, sexual conduct specifically defined by the applicable law; AND
(3) That a reasonable person would find that the work, taken as a whole, lacks serious literary, artistic, political and scientific value.

Chief Justice Burger went on to make clear that 'Under the holdings announced today, no one will be subject to prosecution for the sale or

---

43 Delgado, R, 'Words that wound: A Tort Action for Racial Insults, Epithets and Name Calling', 1982, 17 *Harvard Civil Rights-Civil Liberties Law Review* 133.
44 Vick, D, 'Regulating Hatred' in Klang, M and Murray, A (eds), *Human Rights in the Digital Age*, 2005, London: Glasshouse Press, p 46.
45 This is known as the MacKinnon/Dworkin debate and is found most clearly in the work of Catherine MacKinnon and Andrea Dworkin. See MacKinnon, C, *Feminism Unmodified: Discourses on Life and Law*, 1987, Cambridge, MA: Harvard UP, pp 127–213; Dworkin, A, 'Against the Male Flood: Censorship, Pornography, and Equality', 1985, 8 *Harvard Women's Law Journal* 1.
46 Brigman, W, 'Pornography as Political Expression', 1983, 17 *Journal of Popular Culture* 129. See also Dershowitz, A, 'Op-Ed' *New York Times*, 9 February 1979.
47 *Roth v United States*, 354 US 476 (1957).   48 413 US 15 (1973).

exposure of obscene materials unless these materials depict or describe patently offensive "hard core" sexual conduct specifically defined by the regulating state law, as written or construed."[49] Although at the time it was felt that such a widely drawn standard would lead to wide local differences in obscenity laws this did not turn out to be the case. The scope of community standards was narrowed the next year in *Jenkins v Georgia*,[50] when the Court found that the film *Carnal Knowledge* could not be found to be patently offensive to the local community. Later, further guidance would come in the case of *Pope v Illinois*,[51] which found that the test for literary, artistic, political or scientific value, had to be based upon national, not local, standards. With these cases the Supreme Court put in place an extensive set of legal guidelines with the aim of providing local State and Federal Courts with the tools needed to delineate the boundary between protected expression, which portrayed sexual conduct, and unprotected obscene expression. With the explosion of the internet in the 1990s though, their concepts of local communities with community standards were about to be seriously challenged.

Cyberspace not only removed the barriers between States, it also broke down barriers between communities. As we saw in Chapter 5, as new online communities emerged, individuals started to migrate parts of their lives into the online environment. There, different community standards applied, but, as we have already seen, their actions in the online environment were not without effect in the real world. The US's approach to policing obscenity, much like the UK's, was predicated on the existence of a product fixed in a physical medium and sold or displayed through a physical outlet. The digitisation of pornography rendered the *Miller* concept of 'contemporary community standards', redundant. This first became clear in the case of *US v Thomas*.[52] Robert and Carleen Thomas operated a 'for fee' bulletin board service called *Amateur Action*, which allowed members to download pornographic materials including materials depicting among other things, bestiality, oral sex, incest and sado-masochistic abuse. To view and download these images users had to pay a $55 membership fee and fill out and sign an application that included their name and geographic address. In July 1993, a US postal inspector, David Dirmeyer, received a complaint regarding the Amateur Action BBS from an individual who resided in the Western District of Tennessee. Dirmeyer investigated the complaint and using an assumed name joined the BBS and successfully downloaded content from the Amateur Action BBS to his home computer. As a result of Agent Dirmeyer's investigation, the Thomases were indicted by a grand jury in the Western District of Tennessee and were charged with knowingly using a facility and means of interstate commerce (in this case, the combined computer/telephone system)

---

49 Ibid, p 27.   50 418 US 153 (1974).   51 481 US 497 (1987).
52 *United States v Thomas* 74 F 3d 701 (6th Cir. 1996).

for the purpose of transporting obscene materials. The Thomases argued that they should not be subject to the community standards of Tennessee as they were based in California and in California the materials they posted were not obscene under local community standards (they had purchased the material in local sex shops) and as such they did not and could not know they were committing an offence in Tennessee. Further, they argued that if any community standards other than those of California were to be applied, they should be that of the online community, a community based on cyberspatial rather than geographical connections among people. Both these claims were rejected by both the Distinct Court and the Court of Appeals for the Sixth Circuit, with the Court of Appeals finding that 'obscenity is determined by the standards of the community where the trial takes place' and that 'it is not unconstitutional to subject interstate distributors of obscenity to varying community standards'.[53] Essentially the decision in the *Thomas* case was simplified by the fact that the Thomases knew images hosted on their BBS were being accessed and downloaded in Tennessee (and elsewhere) because they kept a record of subscribers; and in cases where interstate trade occurs with the knowledge of the supplier it has been held to be reasonable to hold them to the standards of that state or community.[54] More worrying cases from the point of view of both lawmakers and parents were, though, about to emerge. In cases where pornographic material was posted onto a publicly accessible BBS or website, it was more difficult to prosecute using local community standards, as in these cases it was not possible to show knowledge or an intent to trade within a particular community. In fact, as cases such as *American Libraries Association et al. v Pataki*[55] and *PSINet v Chapman*[56] were later to demonstrate, attempts to apply local community standards had to be carefully handled lest they be found to be in violation of the implicit confines on State power imposed under the US Constitution's Commerce Clause.[57] In fact during the 1990s, despite the success of a few high-profile prosecutions such as the *Thomas* case, the quantity of online publicly available pornographic content grew rapidly through the development of BBS trading communities, free access websites and fledgling file-sharing systems.[58] With more pornography becoming freely available, US lawmakers were faced with a new problem. At least 'for fee' bulletin board services, such as the one in the *Thomas* case, were largely adult environments; the wider internet was less discriminating and children were able to access all content, including 'adult

53 See above, fn 52, p 710.
54 *US v Beddow* 957 F 2d 1330 (6th Cir. 1992); *US v Williams*, 788 F 2d 1213, 1215 (6th Cir. 1986).
55 969 F Supp 160 (SDNY 1997).   56 63 F 3d 227 (4th Cir. 2004).
57 Art I, s 8, cl 3 of the United States Constitution.
58 Thornburgh, D and Lin, H, *Youth, Pornography, and the Internet*, 2002, Washington DC: National Academies Press, chapter 3.

content', as quickly and easily as adults, for, as noted by Lawrence Lessig, 'a kid in cyberspace need not disclose that he is a kid'.[59]

Faced with a growing problem of children being exposed to online 'adult' content,[60] State and Federal lawmakers attempted to take legal control of the online environment. In 1996, two such attempts came to public prominence, and provoked controversy. In the State of New York, Governor George Pataki oversaw the introduction of §235.21(3) to the New York State Penal Code (NYSPC). This made it a crime to disseminate information 'harmful to minors' via a computer system. At the same time the Federal Government introduced the Communications Decency Act 1996 (CDA) as Title V of the Telecommunications Act of 1996. Both measures were felt to be in breach of the First Amendment by free speech advocates and were immediately challenged. NYSPC §235.21(3) was challenged by an extensive coalition of groups including the American Library Association, Peacefire and the American Civil Liberties Union.[61] They contended that the change in the NYSPC was unconstitutional as it unduly burdened free speech in violation of the First Amendment and it unduly burdened interstate commerce in violation of the Commerce Clause. At a summary hearing on 20 June 1997, the plaintiffs succeeded in their claim and were awarded summary judgment. District Judge Loretta Presky noted that although:

> The State asserted that only a small percentage of Internet communications are 'harmful to minors' and would fall within the proscriptions of the statute . . . I conclude that the range of Internet communications potentially affected by the Act is far broader than the State suggests. I note that in the past, various communities within the United States have found works including *I Know Why the Caged Bird Sings* by Maya Angelou, *Funhouse* by Dean Koontz, *The Adventures of Huckleberry Finn* by Mark Twain, and *The Color Purple* by Alice Walker to be indecent. Even assuming that the Act applies only to pictures, a number of Internet users take advantage of the medium's capabilities to communicate images to one another and, again, I find that the range of images that might subject the communicator to prosecution (or reasonably cause a communicator to fear prosecution) is far broader than

---

59 Lessig, *Code and Other Laws of Cyberspace*, see above, fn 11, p 174.
60 The National Research Council reported that 'one minor in four had at least one inadvertent exposure to sexually explicit images in 1999, with the majority of these exposures occurring to youths 15 years of age or older. The children who inadvertently viewed these images saw them while searching or surfing the Internet (71 per cent), and while opening email, or clicking on links in email or IMs (29 per cent). Most of these exposures (67 per cent) happened at home, but 15 per cent happened at school, and 3 per cent happened in libraries'. Thornburgh and Lin, see above, fn 58, pp 138–9.
61 *American Libraries Association et al v Pataki*, see above, fn 55.

defendants assert. For example, many libraries, museums and academic institutions post art on the Internet that some might conclude was 'harmful to minors'. Famous nude works by Botticelli, Manet, Matisse, Cezanne and others can be found on the Internet. In this regard, I point out that a famous painting by Manet which shows a nude woman having lunch with two fully clothed men was the subject of considerable protest when it first was unveiled in Paris, as many observers believed that it was 'scandalous'. Lesser known artists who post work over the Internet may face an even greater risk of prosecution, because the mantle of respectability that has descended on Manet is not associated with their as yet obscure names. Lile Elam, the founder of Art on the Net, submitted a Declaration that included samples of the types of work found on Art on the Net's site; certain of the images might be considered harmful to minors in some communities, including several nudes and a very dark, disturbing short story entitled 'Two Running Rails of Mercury,' accompanied by a picture of a woman's nude body dissolving into railroad tracks. Rudolf Kinsky testified to his perception of the greater risk run by an unrenowned artist who posts controversial images on the Internet; when he was asked by defendants if a work by Corbet could subject the artist to prosecution, he answered, 'His works are established; they are known. This is a different situation. Could be or could not, but my situation, when I am at the beginning of my career, and someone can, because I am not known, I have no established name and everything, I can still be prosecuted'. Individuals who wish to communicate images that might fall within the Act's proscriptions must thus self-censor or risk prosecution, a Hobson's choice that imposes an unreasonable restriction on interstate commerce.

This is an extremely important passage of an extremely important decision in relation to the legal control of internet content. In this passage Judge Presky sets out the boundaries within which State legislatures must work if they are to produce a set of legal controls, which do not offend against the Commerce Clause, and as we can see, she draws these boundaries narrowly. Although State legislatures retain the power to control the supply of obscene material, a power which the Supreme Court recognised in *Miller v California*,[62] attempts to control the supply of sexually explicit, though not obscene, material are unlikely to be effective given the *Pataki* decision. The problem faced by State legislatures was that they could not sufficiently precisely define the terms of the content they were seeking to control, a problem exacerbated by the lack

---

62 See above, fn 48. Chief Justice Burger made this clear by stating: 'This Court has recognized that the States have a legitimate interest in prohibiting dissemination or exhibition of obscene material when the mode of dissemination carries with it a significant danger of offending the sensibilities of unwilling recipients or of exposure to juveniles.' (p 16).

of a common national standard. As had already been demonstrated by the *Thomas* case, what may be deemed to be acceptable in California, may be felt to be unacceptable in Tennessee, and with State laws requiring individuals to self-censor, it is almost impossible to imagine how such regulations could not offend against the Commerce Clause. What was clearly needed was a Federal response, so was the CDA the answer?

The Communications Decency Act (CDA) was introduced to the Senate on 1 February 1995 by Senators James Exon, a Democrat from Nebraska, and Slade Gorton, a Republican from Washington, in response to the previously discussed fears that internet pornography was on the rise. In March 1995, the Senate Commerce Committee unanimously adopted the Exon/Gorton proposal as an amendment to the in progress Telecommunications Reform Bill. In June 1995, the Senate attached the Exon/Gorton amendment to the Bill by 84 votes to 16. On 1 February 1996, the Bill was passed by both Houses, becoming law on 8 February 1996. The introduction of the CDA explicitly outlawed intentionally communicating 'by computer in or affecting interstate or foreign commerce, to any person the communicator believes has not attained the age of 18 years, any material that, in context, depicts or describes, in terms patently offensive as measured by contemporary community standards, sexual or excretory activities or organs'.[63] Opponents of the Act argued that 'just as a librarian cannot be expected to determine the age and identity of all patrons accessing a particular book in the library's collection, the provider of online information cannot be expected to police the usage of his or her online offerings. To impose such a requirement would result in reducing the content of online material to only that which is suitable for children'.[64] A campaign against the Bill began on its introduction and by 1 February 1996 over 115,000 signatures had been collected on a petition against the Act. On 2 February 1996, in response to the adoption of the Act by Congress, thousands of websites turned black for 48 hours as part of the Electronic Frontier Foundation's (EFF), 'Turn the Web Black' protest. On 8 February 1996, the EFF launched its blue ribbon 'Free Speech Campaign'. This asked those who ran web pages to display a distinctive blue ribbon logo in support of their campaign against the CDA and almost overnight the Blue Ribbon logo populated the web. Publicity campaigns such as these were though merely a sideshow to the main event. As soon as President Clinton signed the CDA on 8 February, the American Civil Liberties Union and 23 other co-plaintiffs, including the Electronic Privacy Information Center, the EFF and the Planned Parenthood Federation of America, raised a complaint

---

63 §. 502(2).
64 Sobel, D, 'The Constitutionality of the Communications Decency Act: Censorship on the Internet' 1 *Journal of Technology Law & Policy*, 1996, 2, available at *http//grove.ufl.edu/~tech law/vol1/sobel.html*.

before the Federal District Court in Philadelphia, seeking a temporary restraining order against the implementation of the indecency provisions of the CDA on the grounds that 'the Act is unconstitutional on its face and as applied because it criminalizes expression that is protected by the First Amendment; it is also impermissibly overbroad and vague; and it is not the least restrictive means of accomplishing any compelling governmental purpose'.[65] The complaint was heard by District Judge Ronald Buckwalter, who, on 15 February, granted the plaintiffs an order, insofar as the CDA referred to 'indecent', but not 'obscene' content.[66] With the order in place the plaintiffs then extracted from the Federal Government a stipulation that they would not 'initiate any investigations or prosecutions for violations of 47 U.S.C. Sec.223(d) for conduct occurring after enactment of this provision until the three-judge Court hears Plaintiffs' Motion for Preliminary Injunction'.[67] With this safeguard in place to ensure that the CDA would not be enforced while a question mark remained over its constitutionality, the plaintiffs prepared a case to be heard before the District Court. Hearings were quickly arranged and held over six days from 21 March to 10 May.[68] The decision was given on 11 June and all three judges agreed that on its face the CDA was unconstitutional. Chief Justice Sloviter reflected the views of the Court in noting: 'I have no hesitancy in concluding that it is likely that plaintiffs will prevail on the merits of their argument that the challenged provisions of the CDA are facially invalid under both the First and Fifth Amendments.'[69] The Federal Government, as expected, immediately sought to appeal the decision to the US Supreme Court, and on 6 December 1996 the Supreme Court noted probable jurisdiction and agreed to hear the case on 19 March 1997. The Government filed its brief on 21 January; the plaintiffs' briefs were filed on 20 February. Oral argument was heard, as scheduled, on 19 March, following which everyone waited for the Court's ruling. The Court finally issued its decision on 26 June, and by a 7–2 majority it found in favour of the plaintiffs. The first decision the Court had to come to was whether the First Amendment applied in cyberspace. Here, Justice Stevens, who gave the majority opinion was clear:

65 Complaint filed before the United States District Court, Eastern District of Pennsylvania, 8 February 1996, Civ. No. 96–963, available at: *www.epic.org/free_speech/censorship/lawsuit/complaint.html*.

66 *ACLU v Reno* Civil Action No. 96–963, available at: *www.epic.org/free_speech/censorship/law suit/TRO.html*.

67 Stipulation of 23 February 1996, available at: *www.epic.org/free_speech/censorship/lawsuit/stipulation.html*.

68 The Plaintiffs' case was heard on 21 and 22 March and 1 April while the Government's case was put on 12 and 15 April. Closing arguments were heard on 10 May.

69 *ACLU v Reno* 929 F Supp 824, p 856.

[The Internet] provides relatively unlimited, low-cost capacity for communication of all kinds. The Government estimates that 'as many as 40 million people use the Internet today, and that figure is expected to grow to 200 million by 1999'. This dynamic, multifaceted category of communication includes not only traditional print and news services, but also audio, video, and still images, as well as interactive, real-time dialogue. Through the use of chat rooms, any person with a phone line can become a town crier with a voice that resonates farther than it could from any soapbox. Through the use of Web pages, mail exploders, and newsgroups, the same individual can become a pamphleteer. As the District Court found, 'the content on the Internet is as diverse as human thought'. We agree with its conclusion that our cases provide no basis for qualifying the level of First Amendment scrutiny that should be applied to this medium.[70]

Thus with the prior question of whether first amendment protection could be applied within cyberspace clearly answered in the affirmative, the Court could go on to assess the constitutionality of the CDA. Again, Justice Stevens was clear:

. . . in order to deny minors access to potentially harmful speech, the CDA effectively suppresses a large amount of speech that adults have a constitutional right to receive and to address to one another. That burden on adult speech is unacceptable if less restrictive alternatives would be at least as effective in achieving the legitimate purpose that the statute was enacted to serve.[71]

The plaintiffs' success was complete. They had won every round and the Supreme Court had, as they hoped, extended First Amendment protection into cyberspace. Congress and the Clinton administration had been roundly defeated, as they would be again in their subsequent attempt to introduce CDA-style legislation, the Child Online Protection Act 1998.[72]

70 *Reno v ACLU* 521 US 844 (1997), p 862.   71 Ibid, p 880.
72 The Child Online Protection Act 1998 attempted a slightly different wording to the CDA by putting more emphasis on knowledge and intent: 'Whoever knowingly and with knowledge of the character of the material, in interstate or foreign commerce by means of the World Wide Web, makes any communication for commercial purposes that is available to any minor and that includes any material that is harmful to minors shall be fined not more than $50,000, imprisoned not more than 6 months, or both.' It was ruled unconstitutional by a 5–4 Supreme Court majority in the case of *Ashcroft v ACLU* 542 US 656 (2004). Available at: *www.mediacoalition.org/legal/copa/supremecourtdecision2004.pdf*.

The decision in *Reno v ACLU* and cyber-regulation

The *Reno* decision was, in Cyberlaw terms, 'the decision heard round the world'.[73] Although the *ratio* of the decision was widely expected, the effects of the decision resonated far beyond the limits of the issue before the Court or even the Supreme Court's jurisdiction. It affected the ability of lawmakers everywhere to use legal controls to manage online content. At its most basic, the *Reno* decision means that 'many of the [Supreme] Court's past First Amendment decisions . . . will be extended to the Internet', including 'application of the strict scrutiny test to invalidate *sub judice* rules prohibiting the publication of information potentially prejudicial to ongoing judicial proceedings . . . and provisions banning hate speech [which the] Court has indicated are unconstitutional'.[74] Thus while the intent of Senators Exon and Gorton may have been to empower courts and law enforcement officials to police cyberspace for the benefit of minors and those who were offended by the wide availability of indecent materials, the effect of their actions proved to be the opposite with all kinds of speech being awarded constitutional protection. Even more far-reaching though was the decision's effect on the ability of courts and law-enforcement authorities to police indecent materials around the globe, for as stated by Douglas Vick, '*Reno's* significance [was] not limited to the territorial boundaries of the United States'.[75] Vick noted that, as Post and Johnson had made clear, material placed on a server anywhere within the US would not only be instantly available across the 50 US States – it would be instantly available worldwide. Material posted on a server based in the US under the protection of the First Amendment could be accessed in the UK in breach of key UK provision such as the Contempt of Court Act 1981,[76] the Official Secrets Acts 1911 and 1989 or the OPA 1959. As Vick explained, '*Reno* limits the options available in the United Kingdom and elsewhere for enforcing policies that would violate the First Amendment if implemented in the United States'.[77] What Douglas Vick, a US trained lawyer and academic living and working in the UK, was trying to say in as gentle a way as possible was that the conjoined effect of the *Reno* decision and the lack of physical borders in cyberspace meant that UK lawmakers, and those of other States, were impotent in those areas protected by the First Amendment. Fortunately, things are not quite as stark as that, as UK law-enforcement agencies were left with three options following the *Reno* decision: (1) they could seek to regulate UK-based consumers rather than overseas suppliers of infringing material; (2) they could negotiate with US

73 This of course is a variant of the famous line from the first stanza of Ralph Waldo Emerson's *Concord Hymn* which usually reads 'And fired the shot heard round the world'. The text is available online at: *www.readbookonline.net/readOnLine/1178/*.
74 Vick, D, (1998) 'The Internet and the First Amendment' 61 MLR 414, p 419.
75 Ibid.   76 In relation to material *sub judice*.   77 Vick, see above n 74, p 419.

lawmakers in the hope of brokering an international settlement to the problem; or (3) they could seek to recreate, artificially, the physical borders that
were previously in place through the application of design controls. The first
option is not really viable in a large, developed, free-market economy such as
is to be found in the UK. One recent survey suggests that 23 per cent of
Britons are installing broadband access to provide, among other advantages,
faster access to pornographic content,[78] and with an unrelated survey putting
the level of broadband penetration in the UK at nearly 10 million households,[79] this suggests there may be as many as 2.3 million UK-based consumers of online pornography. This is substantially higher than the number
of drivers issued with fixed penalty fines for speeding,[80] but whereas fixed
penalty speeding fines are a purely administrative procedure (on the assumption most drivers will not challenge the fixed penalty notice), every case of
online access to pornography would have to be individually evaluated with
reference to the OPA and subsequent case law, before being prosecuted in a
criminal trial. Such an undertaking is simply not feasible and may explain
why, to date, there have been no prosecutions in England & Wales under
either the Customs Consolidation Act or the OPA for importing and privately
viewing obscene material through an internet connection.

The second option (international co-operation) sounds viable and attractive, but like the first is fatally flawed. The flaw again is set out by Douglas Vick:

> The *Reno* decision will constrain the international community's efforts to
> establish a comprehensive body of common rules for regulating Internet
> content. Under American law, treaties and other international accords
> are hierarchically inferior to the provisions of the United States
> Constitution. A treaty provision, just like a congressional statute, is
> unenforceable if it fails to conform with First Amendment law.[81]

This handicap could clearly be seen in negotiations to draft the Council of
Europe's Convention on Cybercrime.[82] The Convention deals with only one
'content-related offence', that being the production or distribution of child
pornography using a computer system.[83] We know several of the States that

78 Orlowski, A, 'One in Four Brits on Net for Porn', *The Register*, 8 December 2004, available
   at: *www.theregister.co.uk/2004/12/08/brit_net_filth/*.
79 Simms, D, 'UK Leads Broadband Stakes in Europe', *ABCMoney.co.uk*, 2 January 2006,
   available at: *www.abcmoney.co.uk/news/0220061643.htm*.
80 Almost 1.8 million speeding offences were detected by UK Speed cameras in 2003. Source:
   The Times, 27 April 2005.
81 Vick, see above, fn 74, p 419.
82 Council of Europe, ETS No. 185, *Convention on Cybercrime*, Budapest, 23 November 2001,
   available at: *http://conventions.coe.int/Treaty/en/Treaties/Html/185.htm*.
83 Art 9.

took part in the drafting process were keen to include further content-related offences, but that these never made the final text. The reason for this is to be found in the Explanatory Report:[84]

> The committee drafting the Convention discussed the possibility of including other content-related offences, such as the distribution of racist propaganda through computer systems. However, the committee was not in a position to reach consensus on the criminalisation of such conduct. While there was significant support in favour of including this as a criminal offence, some delegations expressed strong concern about including such a provision on freedom of expression grounds.[85]

Although the identity of the delegations in question are not revealed, it is clear that at least one of these would be the US delegation who were invited, along with delegations from Japan, Canada and South Africa, to join with the Council of Europe in the project to produce a truly international Cybercrime Convention. The US delegation could not, as Douglas Vick had predicted, sign the US Government up to any treaty provisions that would conflict with First Amendment protection. With child pornography being clearly classed as obscene in US law,[86] Article 9 could be left in place, but any attempts to extend the Convention into race-hate and xenophobic content could not be countenanced by the US delegation because of the principle of the First Amendment. What has followed has been a demonstrable failure of the international community to 'make law' in this area. The Convention States went on to produce an Additional Protocol to the Convention on Cybercrime,[87] which requires signatory States to 'adopt such legislative and other measures as may be necessary to establish as criminal offences under its domestic law, when committed intentionally and without right, distributing, or otherwise making available, racist and xenophobic material to the public through a computer system'[88] as well as criminalising several other related acts, such as making racist threats[89] and insults.[90] The US Department of Justice has indicated that despite being involved in the drafting of the Protocol, 'the United States does not believe that the final version of the Protocol is consistent with its Constitutional guarantees. For that reason, the US has

---

84 Council of Europe, ETS No. 185, *Explanatory Report on the Convention on Cybercrime*, available at: *http://conventions.coe.int/Treaty/EN/Reports/Html/185.htm*.

85 Ibid, para 35.

86 *New York v Ferber* 458 US 747 (1982).

87 Council of Europe, ETS No. 189, *Additional Protocol to the Convention on Cybercrime, Concerning the Criminalisation of Acts of a Racist and Xenophobic Nature Committed Through Computer Systems*, Strasbourg, 28 January 2003, available at: *http://conventions.coe. int/Treaty/en/Treaties/Html/189.htm*.

88 Art 3.   89 Art 4.   90 Art 5.

informed the Council of Europe that it will not become a Party to the Proto-col'.[91] Further, the Department of Justice has made it clear that the US feels it is under no legal duty to assist signatories of the Protocol in enforcing the terms of the Protocol: 'the Protocol is separate from the main Convention. That is, a country that signed and ratified the main Convention, but not the Protocol, would not be bound by the terms of the Protocol. Thus, its author-ities would not be required to assist other countries in investigating activity prohibited by the Protocol.'[92] The US position will fundamentally undermine the effectiveness of the Protocol and its stance has been attacked by European politicians such as Ignasi Guardans, a Spanish MP in the Liberal, Demo-cratic and Reformers' Group, who expressed his frustration by exclaiming: 'If the USA refuses to sign, it must explain to the world why it refuses to co-operate on racism and why it wants to remain a haven for racist websites.'[93] But with the US legally unable to sign the Protocol it serves to remind us how a distinctive set of legal principles in any one nation can undermine the effectiveness of law as a regulatory tool in an international environment without borders.

Thus with international agreement proving impossible in areas where one State, be it the United States or any other State, is constrained by its constitution, there is only one possible option remaining for lawmakers who wish to implement direct legal regulation over the body of cyberspace within their borders, and that is to implement software tools to recreate, artificially, physical borders within the network environment. Such an approach has been widely adopted in those states where content controls are commonplace. A survey by Ronald Diebert and Nart Villeneuve in their book chapter, 'Firewalls and Power: An Overview of Global State Censorship of the Inter-net',[94] reveals that States may implement a variety of filtering and blocking systems including inclusion filtering,[95] exclusion filtering[96] and content

---

91 US Department of Justice, Frequently Asked Questions and Answers: Council of Europe Convention on Cybercrime, available at: *www.usdoj.gov/criminal/Cybercrime/COEFAQs.htm*.
92 Ibid.
93 Council of Europe, Interview with Ignasi Guardans, 23–27 September 2002, available at: *www.coe.int/T/E/Com/Files/PA-Sessions/Sept–2002/Int_Guardans.asp*. Also of interest in rela-tion to this issue are the *UEJF and Licra v Yahoo! Inc. and Yahoo France* cases. Tribunal de Grande Instance de Paris, 22 May 2000 (*www.juriscom.net/txt/jurisfr/cti/yauctions 20000522. htm*); Tribunal de Grande Instance de Paris, 20 November 2000 (*www.lapres. net/yahen11. html*); *Yahoo Inc v LICRA*, 7 November 2001, District Court (ND Calif) (*www.cdt. org/jurisdiction/011107judgement.pdf*); *Yahoo Inc v LICRA*, 23 August 2004, Court of Appeals for the 9th Circuit (*www.ca9.uscourts.gov/ca9/newopinions.nsf/D079531C495BC5 E288256EF90055E54C/$file/0117424.pdf?openelement*).
94 Diebert and Villeneuve, see above, fn 41.
95 Inclusion filtering allows users access a short list of approved sites, known as a 'white list', only. All other content is blocked. Ibid, p 112.
96 Exclusion filtering restricts user access by blocking sites listed on a 'black list'. All other content is allowed. Ibid.

analysis[97] to block access to certain content from within their borders. Many States have chosen to implement these systems to provide for some form of control over their citizens. France and Jordan for instance have chosen to implement 'limited' controls,[98] with the Tribunal de Grande Instance de Paris ordering Yahoo! to block access to auction sites offering Nazi memorabilia,[99] and with Jordan blocking access to one website at a national level.[100] Others such as Vietnam and the US have implemented distributed controls.[101] Vietnamese ISPs use proxy servers with an access control list to block access to many news, human rights and dissident websites, while schools in the US are required under the Children's Internet Protection Act to implement 'Internet safety measures'. The most invasive and wide-ranging controls, known as comprehensive controls,[102] are used by 12 States with a long history of media controls and censorship.[103] These comprehensive controls are the only controls that seek to completely block access to a variety of websites from within the State. It is controls of this order that individual States such as the UK would need to put in place to control their borders completely, but the costs of such an action in personnel, economic and political terms should not be underestimated. Although exact figures are almost impossible to source, a report in 2003 suggested that the Chinese Government employs 30,000 full-time staff to manage the state firewall,[104] a number which almost exactly matches the current size of the Metropolitan Police Force.[105] In addition to the high fiscal cost of employing these staff, the firewall has a highly deleterious effect on the Chinese economy. Ken DeWoskin, a partner in PricewaterhouseCoopers commented that 'You have a lot of talent, not to mention money, that is being directed into controlling rather than stimulating the use of the Web . . . it's like an enormous tax in terms of time and cost that is introduced into the use of the internet for research . . . everything is just slow as molasses'.[106] China can absorb these high personnel, fiscal and

---

97   Content analysis restricts user access by dynamically analysing the content of a site and blocking sites that contain forbidden keywords, graphics or other specified criteria. Ibid.

98   Limited controls means access is restricted to a small number of websites. Ibid, p 121.

99   *UEJF and Licra v Yahoo! Inc. and Yahoo France*, see above, fn 93.

100 The website in question is *www.arabtimes.com*.

101 Distributed controls mean access is restricted to a significant number of sites, but sporadically implemented by different ISPs. Diebert and Villeneuve, see above, fn 41.

102 Comprehensive controls mean that access is restricted to a number of sites within a comprehensive national framework. Ibid.

103 The twelve are: Bahrain, China, Cuba, Kazakhstan, Myanmar, Saudi Arabia, Singapore, Syria, Tunisia, United Arab Emirates, Uzbekistan and Yemen.

104 Source: BBC News, *The Cost of China's Web Censors*, available at: *http://news.bbc.co.uk/2/hi/business/2264508.stm*.

105 In 2005 there were 30,235 full time police officers of the Metropolitan Police Service. Source: *www.met.police.uk/about/*.

106 BBC News, see above, fn 104.

economic costs due to its large size and rapid economic growth. Equally the People's Republic of China has little concern for the political costs of implementing such extensive controls as there is no likelihood of the Government being voted out of power and civil rights activists may be censored by the apparatus of state. But in a western free-market democracy these costs are a strong disincentive to any lawmaker who sought to implement the comprehensive control model. There is little likelihood that the UK Government could afford the investment required to 'rebuild our borders', and even if the political will were there, such a plan is unlikely to be accepted by the people, or the media, of the UK. The truth is that for most democratic States the option of building 'digital borders' is simply not available, due to the political costs involved.

Laws in cyberspace

The experience gathered from these failed attempts to regulate indecent and obscene content in cyberspace, serves as a warning to lawmakers hoping to directly regulate the actions of netizens through the application of legal regulation. Indecency was one of the earliest aspects of net content and culture which came to the attention of lawmakers, and yet 10 years on, no country has successfully eradicated access to obscene material hosted in cyberspace from within its borders without first building a comprehensive filtering and blocking network. All similar attempts to control access to hate speech and racially offensive content have been equally unsuccessful. Both Germany and France have attempted to block access to pro-Nazi discussion boards and websites, but have met with little success. In 1996, state prosecutors in Mannheim used the threat of litigation to force Germany's largest ISP, T-Online, to block access to a California-based neo-Nazi website operated by Ernst Zündel, a German citizen living in Canada.[107] The action, although immediately effective, was soon undermined. Free-speech advocate Declan McCullagh, although not endorsing Mr Zündel's views, demonstrated to the Mannheim prosecutor the futility of trying to censor speech on the internet. He mirrored Zündel's website to computers housed at Stanford, Carnegie Mellon University and MIT, in effect daring the German authorities to cripple German research by shutting off access to these major academic servers. Other free-speech advocates used his 'Zundelsite kit' to scatter mirror sites across the Net.[108] David Jones, President of Electronic Frontier Canada, noted that rather than having the desired effect, the actions of the Mannheim

---

107 Nash, N, 'Germany Moves Again to Censor Internet Content', *New York Times*, 29 January 1996.

108 The 'Zundelsite kit' was a single file created by McCullagh, containing all the content of Zündel's website posted in a message to a Usenet newsgroup with instructions on 'how to open your very own Zundelsite mirror archive in five minutes or less'.

prosecutor led to 'the information being copied to new locations in Cyberspace and becoming even more accessible . . . and with the publicity, more people might want to visit these web pages to see what all the fuss is about'.[109] Similarly, the efforts of the Tribunal de Grande Instance de Paris to block access by French citizens to Yahoo! auction sites offering Nazi memorabilia only received some measure of success, thanks to a voluntary decision on the part of Yahoo! to cease hosting such material, a decision that the company appears to have taken for public relations reasons. In fact a careful examination of the fallout of the *UEJF and Licra v Yahoo!* litigation reveals that it was community pressure, rather than the legal order, which forced Yahoo!'s hand. Senior auction producer Brian Fitzgerald in an interview with *The Guardian* newspaper said that the court order played no role in the new policy other than to raise awareness internally and speed the decision. Rather, he said, that while some users support the trade of such items on free-speech grounds, the majority of comments received by Yahoo! were in opposition, and this was why Yahoo! decided to change its listing policy.[110] While sceptics may suggest this is simply a face-saving response from Yahoo! the subsequent decision in the case of *Yahoo Inc v LICRA*,[111] a decision described by Michael Geist of the University of Ottawa, as 'a no brainer',[112] suggested there was no immediate legal imperative for Yahoo! to take such an action.

The failure of legal controls to regulate in all these cases could be predicted by applying the principles described in Johnson & Post's paper, *Law and Borders – The Rise of Law in Cyberspace*.[113] While some cyberpaternalists may say that failure to regulate internet content in certain cases does not equate to an inability to use legal controls to regulate in cyberspace, I believe that direct legal regulation will always fail in such circumstances. While supporters of direct legal intervention may suggest that the fact that the Convention on Cybercrime has produced international agreement to combat activities such as hacking,[114] virus writing and seeding,[115] and denial of service,[116] is evidence of the ability of direct legal controls to flourish in cyberspace, I would suggest this is not likely to produce an effective legal response to these problems. While the Convention will see international co-operation

---

109   Electronic Frontier Canada, Press Release, 1 February 1996. Full text at Braun, F, Gerlach, C and Münch-Dalstein, M, *Rechtsprobleme des Internet: Regulierung und Selbstregulierung*, Wintersemester, Projektgruppe an der Juristischen Fakultät der Humboldt-Universität, p 44. Available at: *www.muench-dalstein.de/rdi/rdi.pdf*.

110   'Yahoo! to Stop Auctions of Nazi Memorabilia', *The Guardian*, 3 January 2001. Available at: *www.guardian.co.uk/internetnews/story/0,7369,417384,00.html*.

111   District Court (ND Calif), see above, fn 92.

112   Quoted in Kaplan, C, 'Cyber Law Journal: Was the French Ruling on Yahoo! Such a Victory After all?' *New York Times*, 16 November 2001, available at: *http://Cyber.law.harvard.edu/is02/readings/nyt-yahoo-2.html*.

113   See above, fn 3.   114   Art 2.   115   Art 4.   116   Art 5.

among law-enforcement officials of the 50 Convention States,[117] it will not provide effective regulation of these activities for as long as one nation remains outside the Convention. As I hope I have demonstrated in this chapter, to be completely effective, legal controls in cyberspace must be supported by *all* countries and governments, otherwise we will have a scenario similar to the ones we examined in our case studies where there will always be an international 'safe harbour' for those wishing to carry out particular forms of antisocial behaviour, whether it be the distribution of pornography, the peddling of hate speech, or the sending of unsolicited 'spam' messages. To control these actions without the assistance of these safe harbour States will involve law-enforcement officials making similar choices to the ones discussed in relation to content. They will have to either (1) invest large amounts of both time and money to identify and prosecute their citizens under domestic laws; or (2) seek to negotiate with 'Rogue States' that provide these safe harbours in the hope of brokering an international settlement to the problem; or (3) seek to recreate, artificially, the physical borders that were previously in place through the application of design controls. In truth, the only practical option – in most cases is the first option – and with the scale of investment that is required, law-enforcement agencies tend to focus their actions on only the most high-profile cases such as that of Jeremy Jaynes[118] or Kevin Mitnik.[119] Most amateur hackers and script-kiddies will continue to be able to carry out their antisocial activities without fear of detection or prosecution, whether or not their home state has signed up to the Convention on Cybercrime.[120]

What role does this leave for lawmakers? Despite the pessimistic tone that is adopted in this chapter, traditional lawmakers still have a key role to play in the new cyber-regulatory environment. As we saw in Chapter 2, the key is to recognise that legal-regulatory controls should not be expected to be directly effective. Instead, one should seek to develop a regulatory hybrid control system in which legal controls form part of the regulatory web or matrix. It is to this we turn in the next chapter.

---

117 This is assuming all countries implement the convention, which seems quite unlikely as to date only 11 have.
118 According to the Register of Known Spam Operations database, Jaynes was the world's eighth most prolific spammer who had defrauded thousands of individuals and businesses. He was prosecuted in Louden County, Virginia in April 2005.
119 Mitnik is probably the world's best known hacker. He was prosecuted in 1995 following a 15-year hacking career, which saw him steal thousands of programs and data files and over 20,000 credit card numbers from computer systems all over the US.
120 Script kiddies is a derogatory term for inexperienced crackers who use scripts and programs developed by others for the purpose of compromising computer accounts and files, and for launching attacks on whole computer systems. In general, they do not have the ability to write said programs on their own.

Part III

Regulating cyberspace: Challenges and opportunities

Chapter 8

Regulating cyberspace

The Internet is the first thing that humanity has built that humanity doesn't understand, the largest experiment in anarchy that we have ever had.

Eric Schmidt

The case studies contained in Chapters 4–7 demonstrate an interesting pattern. Attempts to externally impose a regulatory settlement, such as the efforts of software companies to leverage control into the content layer via their proprietary software suites, the actions of ICANN with regard to domain names, or public sector attempts to control the distribution of content through legal tools such as the Digital Millennium Copyright Act (DMCA) or the Communications Decency Act (CDA), often lead to unexpected outcomes and ultimately, failure to achieve the regulatory settlement that was sought. Alternatively, on the few occasions where a regulatory settlement has been allowed to develop organically, such as the development of network protocols including TCP/IP and FTP, the community response to the Live 8 ticket affair or the market response to the problem of the home video recorder, a strong and coherent regulatory outcome has developed. What can regulators, and regulatory theorists, learn from these outcomes? Does it mean that direct regulatory interventions will always produce an unplanned and often undesirable outcome? In turn does this suggest a belated victory for the cyber-libertarian school? Certainly it seems to suggest that the concern expressed by the cyberpaternalists that code writers may effect perfect control through software design is unfounded. Although on the surface these seem to be reasonable conclusions to be drawn from the available evidence, the outcome is more complex than all these suggestions, but it is one that offers regulators and regulatory theorists an opportunity they must seize.

Complexity in regulatory design

Regulators must first accept the challenge of regulating in a complex regulatory environment. In Chapter 2 we discussed and analysed the complexity of the regulatory environment and noted several key aspects. The first was that all actors in the regulatory environment play an active role in that environment. You cannot differentiate between regulators and regulatees as all actors act in both roles simultaneously and concurrently. Thus the nominal 'regulatee' will, by reference to their internal value set, choose to accept or challenge any changes that occur within the regulatory environment; as such they play a key role in the development of effective regulatory settlements.[1] Next we acknowledged that while in the physical world regulators may choose to adopt one of several socially mediated regulatory tools or hybrids, which include legal interventions, market controls and social norms, or they may choose to invest in design-based controls to harness the environmental regulation that surrounds us: in cyberspace the issue is further complicated by the plasticity of the environment of cyberspace.[2] This is additionally complicated when dealing with media regulation because of the effect of 'layering', which is caused by using carrier layers to 'fix' and distribute media content. An analysis of layering revealed the effective vertical regulation could transfer from the supporting carrier layers to the higher content layer.[3] Finally, by examining simple regulatory scenarios, such as content control within the People's Republic of China (PRC), we see that the number of competing regulatory demands leads to a high degree of regulatory competition and a complex three-dimensional regulatory matrix developed (replicated below in Figure 8.1).[4]

At the time, we noted that 'at each point in the matrix, a regulatory intervention may be made, but the complexity of the matrix means that it is impossible to predict the response of any other point in the matrix'.[5] Now having carried out extensive case studies, and having isolated some successful and some unsuccessful attempts to intervene in the complex regulatory matrix found in cyberspace, what may we learn from this model?

The value of the regulatory matrix, for regulators and for regulatory theorists, is as a substitute for traditional static regulatory models.[6] The case studies in Chapters 4–7 allow us to map when a regulatory intervention is likely to be successful and when it will be unsuccessful. If we look at the failure of ICANN to achieve widespread acceptance within the cybercommunity, and with it legitimacy, we see structural failures in the regulatory intervention that led to ICANN's creation – in other words ICANN was flawed from its inception. If we map the regulatory matrix surrounding the

---

1   See above, pp 24–5.   2   See above, pp 37–43.   3   See above, pp 43–6.
4   See above, pp 47–54.   5   See above, p 53.   6   See above, pp 27–30.

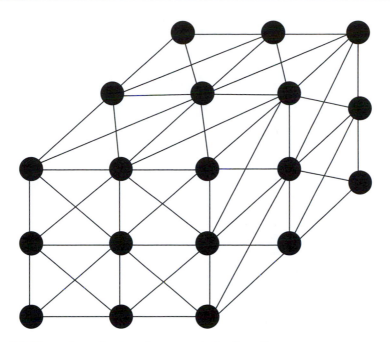

Figure 8.1 Three-dimensional regulatory matrix (replicated).

creation of ICANN (as is seen in Figure 8.2) we can immediately see the sources of regulatory tension that undermined ICANN.

ICANN was created by an executive action of the US Government.[7] This action – represented by Point A in Figure 8.2 was an external regulatory intervention into the settled regulatory matrix. It was the intention of the US Government to bring stability to the process of awarding and managing domain names and to bring a degree of public accountability to the process. In fact the existence of ICANN has arguably destabilised the domain name system, while ICANN itself has been repeatedly criticised for being unaccountable.[8] The question this raises for regulators and regulatory theorists is why has this happened? Fortunately, some of the reasons for ICANN's regulatory failures become apparent when we examine the effect it had on the regulatory matrix. Point B represents the UN in the guise of the World Intellectual Property Organisation (WIPO). WIPO saw the creation of ICANN initially as a threat, then as an opportunity. When invited by the US Department of Commerce (DOC) to create a set of policy recommendations

7   See above, pp 104–7.
8   Mueller, M, *Ruling the Root*, 2002, Cambridge, MA: MIT Press; Froomkin, M, 'Wrong Turn in Cyberspace: Using ICANN to Route Around the APA and the Constitution', 2000, 50 *Duke Law Journal*, 17.

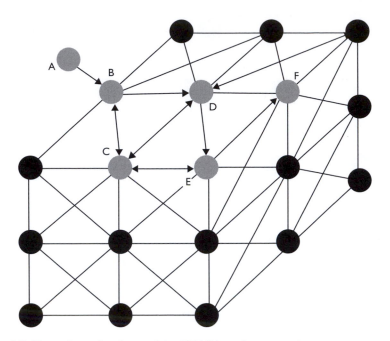

Figure 8.2 Three-dimensional map of the ICANN regulatory matrix.

for ICANN with regard to intellectual property rights (IPRs), WIPO produced first a Green Paper,[9] then a Final Report,[10] highly favourable to trade mark holders. In so doing WIPO created further changes, and tensions, within the regulatory matrix. One was the effect of alienating a large proportion of domain name owners, represented by Point C. Critics claimed ICANN was biased in favour of trade mark holders,[11] with the community responding both through organised media campaigns[12] and more directly through the election of highly critical candidates in ICANN's At-Large elections.[13] The actions of regulatory bodies such as the US Government and WIPO, not only affected consumers: regulatory tensions were also created with other regulators. The EU, represented at Point D, was concerned that the creation

9 WIPO, *The Management of Internet Names and Addresses: Intellectual Property Issues*, RFC–3, 23 December 1998, available at *http://arbiter.wipo.int/processes/process1/rfc/3/ interim2.html*. Discussed above, pp 110–13.

10 WIPO, *The Management of Internet Names and Addresses: Intellectual Property Issue, Final Report of the WIPO Internet Domain Name Process*, 30 April 1999, available from *http:// arbiter.wipo.int/processes/process1/report/pdf/report.pdf*. Discussed above, pp 111–12.

11 Mueller, see above fn 8; Murray, A, 'Regulation and Rights in Networked Space', 2003, 30 JLS 187.

12 *www.icannwatch.org/*; *www.internetgovernance.org/* and *http://at-large.blogspot.com/*.

13 See discussion above, pp 114–18.

of ICANN could establish 'permanent US jurisdiction over the Internet as a whole, including dispute resolution and trademarks used on the Internet'.[14] Although some of the concerns of the EU were addressed by the US DOC, there remains a degree of tension between the EU and ICANN, which permeated the extensive discussions on the creation of the .eu top level domain. The EU's actions did not, though, end with the creation of ICANN. The EU States are, of course, powerful members of the UN, and they, along with many others, have pushed the issue of cyber-regulation onto the UN agenda through the World Summit on the Information Society (WSIS), which is represented in our model by Point E.[15] The WSIS process has, to date, provided the most wide-ranging review of internet governance in general and the role of ICANN in particular. Although predictions that the WSIS process would eventually lead to the extinction of ICANN to be replaced by a 'truly international' regulatory body,[16] have thus far proved fruitless, there is no doubt that the creation of the new Internet Governance Forum (IGF) (Point F) will continue the work of WSIS. It provides a forum for governments such as China, Cuba and South Africa to continue to press their case for the reshaping of regulatory bodies such as ICANN and as such the IGF becomes a key player in the regulatory matrix. Whatever results eventually come from the WSIS/IGF process, we can confidently predict that they will create further regulatory tensions throughout the regulatory matrix and are unlikely to solve the current problems of ICANN and the domain name system. Thus, by simply modelling ICANN's failings we can predict that attempts to impose an unsympathetic regulatory settlement are likely to lead to unplanned tensions and turmoil within the regulatory matrix, undermining the effectiveness of the regulatory intervention. A new ICANN is unlikely to have any more success than the old.

By accepting that the regulatory matrix is a dynamic structure, regulators and regulatory theorists are offered the opportunity to produce effective *complimentary* regulation. We can see the value of such regulation if we look at the position the film industry reached by default after the failure of the *Sony* litigation in 1984[17] or the position the music industry is (slowly) moving toward in the digital music market. In both cases a market led solution provides the most effective regulatory settlement. If we use as our case study the effects of the video cassette recorder (VCR) on the film

---

14 Council of the European Union/European Commission, *Reply of the European Community and its Member States to the US Green Paper*, March 1998, available at *http://europa.eu.int/ ISPO/eif/InternetPoliciesSite/InternetGovernance/MainDocuments/ReplytoUSGreenPaper .html*.

15 See discussion above, pp 118–24.

16 Eg Murphy, K, 'Who Really Runs the Internet?', *Computer Business Review Online*, 14 October 2005.

17 *Sony Corp of America v Universal City Studios* 464 US 417 (1984).

industry in the 1980s and 1990s we see the value of *complimentary*, or *symbiotic* regulation'.[18] If we map the regulatory matrix surrounding the development of the VCR, post 1984 (as is seen in Figure 8.3) we see why it was not the Boston Strangler of the film industry, but rather its Fairy Godmother.[19] What we note first is the doomed attempt of the film industry to externally regulate the technology of the VCR in the failed *Sony* litigation. This is represented at Point A and it should be particularly noted that with the failure of this action the external forces on the regulatory matrix are shifted causing the regulatory focal point to shift from Point A, as was the case in the ICANN case study, to Point C. As with Point C in the ICANN study, Point C here represents the consumers, who, freed from the external constraints of hierarchical intervention took the lead in designing market-led regulatory developments. The consumers immediately began to transmit their demands to the other key players in the VCR marketplace: the hardware suppliers, represented at Point B; the content suppliers, represented at Point D; and movie theatres, represented at Point E.

Consumers demanded from hardware suppliers ever better picture and sound quality, longer playing tapes and easy-to-use recording systems that

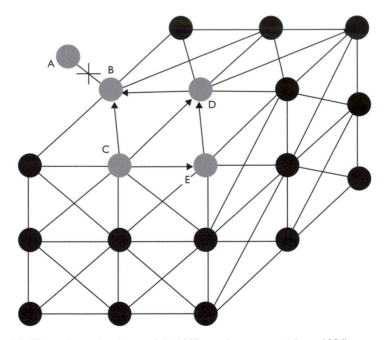

Figure 8.3 Three-dimensional map of the VCR regulatory matrix (post 1984).

18 See also discussion above, pp 201–2. A discussion of 'symbiotic regulation' follows.
19 See above, p 202.

would allow them to programme their VCR for days and weeks ahead. As we moved from the analogue to the digital, consumers demanded better quality and greater storage offerd by digital media such as DVDs. The industry has responded by producing higher quality home video equipment at ever lower prices[20] and has been rewarded by growing consumer expenditure on home entertainment products. Consumers indicated to the movie industry that they were willing to pay for a copy of their favourite movie, which they could watch at home over and over in a fashion similar to playing their favourite record again and again. Further, they indicated they would be willing to pay more for added extras, which were made available through special editions or re-mastered originals. As a result the market for prerecorded videos (and later DVDs) exploded.[21] The video rental market, as exemplified by the success of the Blockbuster chain, offered a whole new market segment, the opportunity to watch recently released movies in the comfort of the consumer's own home before they became available on general sale, but after their theatrical run, brought on tap a whole new income stream for the film industry. This innovation also allowed consumers to bring pressure to bear on the cinema chains, who for many years had been underinvesting in their theatres. Faced with the threat of the clean, well-lit and family-friendly Blockbuster chain, cinema operators invested heavily in their infrastructure throughout the 1980s, leading to the development of the modern multiplex cinema and with it a renaissance in the movie theatre industry.[22] The result of this consumer-led market-regulatory settlement has been success for all parties: consumers have greater choice and higher quality home cinema experiences; home electronics suppliers have new market segments to exploit; the film industry is making increased profits, both at the cinema and through the development of a new market segment: the sale-through video and even the movie theatre industry has benefited from the halo effect, and from increased investment with more customers coming through their doors to see blockbuster spectaculars such as the *Lord of the Rings* Trilogy, the *Spiderman* movies and the *Harry Potter* movies.

What is the key difference between the ICANN case study and the VCR case study, which leads to such a dramatic difference in outcome? It is simply

20 In 1984 a VCR could cost on average between $400 and $500. In 2005 a DVD-R could be bought for under $150.

21 Figures from the UK Film Council reveal that *in the UK alone* in 2004, 153 million VHS videos and DVDs were rented, while 234 million VHS videos and DVDs were sold with a combined market value of £3.13 billion (more than three times the UK theatrical market), see *www.ukfilmcouncil.org.uk/statistics/yearbook/?y=2004&c=10*.

22 Statistics provided by the Film Distributors' Association reveal that in 1984 cinema admissions had fallen to an all-time low of 54 million admissions in the UK (down from 1,635 million in 1946). Since then admission figures showed a steady improvement to reach 165 million in 2005 a figure in excess of that achieved in the years 1972–1980, before the widespread distribution of the home VCR in the UK.

that in the ICANN case study an attempt was made to engineer a regulatory outcome by directive, external, intervention – an intervention that was designed with little regard for the relationships between actors in the extant regulatory matrix. In the VCR case study, fortunately for all involved, an attempt at a similar action ultimately failed and in its place a regulatory settlement evolved organically from within the extant regulatory matrix. It is a lesson that should not be lost on regulators and regulatory theorists. By acknowledging the complexity of the extant regulatory environment, and by developing a dynamic regulatory model we can design more effective regulatory interventions – interventions that take account of the extant regulatory matrix and are more likely to achieve the desired regulatory outcome. Regulators may thus learn from, and apply, the mathematical model of the Gardener's Dilemma. Complex systems may prove to be mathematically intractable but this does not mean that they are unregulated: attempts to intervene in the extant regulatory settlement are, applying Chaos Theory, more likely to disturb the regulatory settlement in an unexpected and unpredictable manner than to achieve the desired outcome, whereas modelling and harnessing the extant regulatory settlement in a dynamic regulatory matrix allows regulators to harness the regulatory relationships already in place. It is the difference between a disruptive regulatory intervention and complimentary intervention, and is the key to successful regulation, both in cyberspace and in real space.

Effective regulation: Disruptive and symbiotic regulation

Disruptive regulation is a term not often used by regulatory theorists who instead prefer to use the term regulatory failure.[23] In using such language though they direct themselves to examine the outcome of a regulatory intervention rather than its effect on the regulatory environment. By this I mean they remain wedded to the static regulatory model and assume that the intervention made by the regulator was the only change made to the extant regulatory environment: then in the event of regulatory failure they examine why *that* regulatory intervention failed rather than examining the polycentric effects of the intervention within the regulatory matrix. This may be a particular malaise of those who follow a traditional command and control approach to measuring regulation, for when' we extend the scope of our

---

23 Eg Joskow, P, 'Regulatory Failure, Regulatory Reform, and Structural Change in the Electric Power Industry', 1989, *Brookings Papers on Economic Activity: Special Issue* 125; Barth, J, Trimbath, S and Yago, G (eds), *The Savings and Loan Crisis: Lessons from a Regulatory Failure*, 2004, New York: Springer-Verlag; Lubulwa, A, *The Implications of Regulatory Failure for Rail and Road Industries*, 1990, London: Avebury Press.

examination to include all forms of regulation, including the regulatory effects of markets and technological innovation, we see a shift in the language. For example in the study of markets the concept of 'creative destruction', developed by the Austrian economist Joseph Schumpeter,[24] plays a key role in our contemporary appreciation of the function of markets. Schumpeter's thesis draws upon analogies from biology and evolutionary theory and suggests that innovations in markets and in technology 'revolutionize the economic structure *from within*, incessantly destroying the old one, incessantly creating a new one. This process of Creative Destruction is the essential fact about capitalism. It is what capitalism consists in and what every capitalist concern has got to live in'.[25] This concept, which revolutionised the study of economics, led to Schumpeter being labelled the most important economist of the twentieth century by the Wall Street Journal, is the current focus of intense debate in the fields of economics and finance,[26] and has been applied to, among other things, the study of the evolution of digital media[27] and new technologies.[28] Thus we examine the process of disruptive developments in some regulatory modalities, but not it seems in others, and not, importantly, when studying and examining the process of regulation as a whole.

One benefit of acknowledging the complexity of the regulatory environment, and moving from a static regulatory model to the dynamic regulatory matrix suggested in this book, is that it allows us to map the effects of a Schumpterian-style disruptive regulatory intervention rather than just its outcome. It is an intervention, such as this, which is measured at pages 234–237, where we examined the failure of the ICANN regulatory intervention, and it similarly describes the intervention of the media distribution industries into the market for digital products and their distribution, described and discussed at length in Chapter 6. As we saw in both cases the disruptive effects of such intervention were not merely centred upon the focal point of the regulatory intervention. In each case the intervention made by the

---

24 *Capitalism, Socialism and Democracy*, 1994 (originally published 1942) Routledge: London.
25 Ibid, p 82.
26 Eg Metcalfe, J (ed), *Evolutionary Economics and Creative Destruction*, 1998, London: Routledge; Caballero, R and Hammour, M, 'On the Timing and Efficiency of Creative Destruction', 1996, 61 *Quarterly Journal of Economics* 805.
27 Eg Dowling, M, Lechner, C and Thielmann, B, 'Convergence: Innovation and Change of Market Structures between Television and Online Services', 1998, 8 *International Journal of Electronic Markets* 31; Ku, R, 'The Creative Destruction of Copyright. Napster and the New Economics of Digital Technology' 2002, 69 U Chi L Rev 263; DiMaggio, P *et al.*, 'Social Implications of the Internet', 2001, 27 *Annual Review of Sociology* 307.
28 Eg Evans, N, *Business Innovation and Disruptive Technology*, 2002, Paramus, NJ: Prentice Hall; Bower, J and Christensen, C, 'Disruptive Technologies: Catching the Wave' *Harvard Business Review* (January-February) 1995, 43; Foster, R, ven Beneden, P and Kaplan, S, *Creative Destruction: Turning Built-to-Last into Built-to-Perform*, 2001, Paramus, NJ: Prentice Hall.

regulatory agency had unforeseen effects on the surrounding nodes of the regulatory matrix. In relation to ICANN these effects were mapped in Figure 8.2 on page 236, where we saw a series of interrelated transmissions take place between competing regulatory bodies in question. If we carry out a similar mapping exercise for the regulation of copyright materials in the online environment, we find a rather more complex model develops with many more interested parties reacting to the actions of the media distribution industries and as a result far greater regulatory uncertainty.

Figure 8.4 is a vastly simplified model of the regulatory matrix that surrounds the management of digital media files in the online environment. In this model, the media distribution industries, represented at Points A, make several discrete attempts to influence the regulatory matrix, including lobbying the governments of the US (Point B) and EU (Point F) to make legislative changes designed to give the media distribution industries greater control over such content in the digital environment. They also use design-based controls such as Digital Rights Management (DRM) systems, represented at Point H, to attempt to 'lock-up' content with the same aim in mind. Unfortunately the peer-to-peer (P2P) providers (represented at point C) initially responded negatively to such action, and sought to relocate their services outside the direct control of the State-based regulators and in so doing challenged their ability to regulate such technology. Consumers, represented

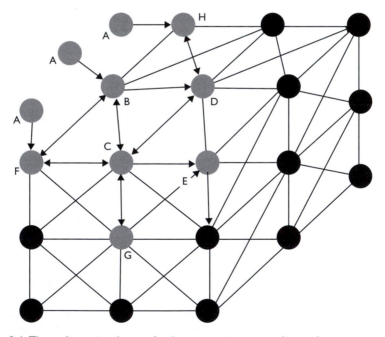

Figure 8.4 Three-dimensional map of online entertainment media regulation.

at Point D, have responded by unlocking digital content, and by creating new pressure groups such as Creative Commons (represented at point E). Finally, Point G represents the next generation of P2P, or similar technology, which will no doubt rise phoenix-like from the ashes of the *MGM v Grokster* litigation.[29] There is now little chance the media entertainment industries can take control of the online market in digital creative goods, using command and control regulation. As the Gardener's Dilemma illustrates, once you start to make changes to the regulatory environment you cannot model how the changes you put in place will effect that environment. This is not to say that the situation remains irretrievable. The success of commercial online distribution providers such as iTunes demonstrates that an eventual regulatory settlement may still be attained by listening to the requirements of the other actors within the regulatory matrix, and by communicating to them the needs of the industry. In so doing all the actors in the matrix may eventually agree a new regulatory settlement (in part the VCR case study proves this). This though is a protracted and potentially damaging process for all parties as it requires them to pass through a lengthy period of regulatory uncertainty before the settlement emerges. This complex process, labelled 'disruptive innovation' by the UK computer theorist John Naughton,[30] takes a considerable time to complete and while the market may be able to accommodate this uncertainty through the use of premium payments to meet the risk such uncertainty brings, the regulatory matrix as a whole is destabilised by such uncertainty. This causes an apparent regulatory vacuum which, as we have already seen, regulators of all forms rush to fill through traditional command and control techniques.[31] Much better is to design a control model that affords all participants in the regulatory matrix an opportunity to shape the evolutionary development of their environment; evolution rather than revolution is the key to effective regulatory intervention and this means that communication between all parties is essential.

According to the dynamic regulatory matrix the best regulatory model is not one built upon an active intervention into the settled regulatory environment, the result of which is likely to be extremely disruptive, rather it is one

---

29 *MGM et al v Grokster et al* 125 S.Ct. 2764 (2005).

30 Naughton differentiates between incremental innovation, practised by established businesses from within a 'regulatory cocoon'; this sees slow development by increments such as from 8-track to compact cassette and disruptive innovation, which is usually practised by new entrants into a market that disrupts the regulatory matrix, and in particular the market, for a period of time. Examples of disruptive innovation include the iPod and the World Wide Web. See Naughton, J, 'How Apple Saved the Music Biz', *The Observer* 13 February 2005, available at *http://observer.guardian.co.uk/business/story/0,6903,1411672,00.html*.

31 For instance there is no reason to assume that the media distribution industries intend to abandon their 'command & control' policies such as lobbying for extensive enforcement powers, the implementation of DRM or their policy of pursuing infringing end-users in the near future.

that harnesses, as best as possible,[32] the relationships already in place between the actors: what I will call *symbiotic regulation*. The development of symbiotic regulation, although complex, is not impossible. It is used in community-led and market-led regulatory developments such as those we studied in Chapters 4–6: in particular the development of network protocols and the VHS/DVD market. How, though, can hierarchical regulators, who are used to implementing a command and control model, match the complexity of these organic regulatory developments? The answer may be to use con-temporary modelling techniques to predict where tensions will arise within the regulatory matrix and to design a regulatory intervention to avoid such tensions and to harness instead the natural communications flows within the matrix: in other words to mimic organic regulatory developments. To do this the regulator must carry out a two stage evaluation process before they start to design their intervention. The first stage is to map the communications that naturally occur between regulatory actors and the second is to predict what feedback will occur after the intervention is made. The first requires them to take account of Niklas Luhmann's theories of autopoietic social systems, the second requires them to be familiar with system dynamics.

Modelling symbiotic regulation: Autopoiesis and systems dynamics

Niklas Luhmann's thesis of autopoiesis[33] develops Humberto Maturana and Francisco Varela's biological concept of autonomous living systems[34] and proposes that social systems are self-referring entities, created within their own organisational logic. This approach is a radical departure from mainstream sociological thought, which is based on the premise of collective human agency. According to Luhmann there is no central organisational body and no hierarchical structure; merely unique subsystems, and sub-systems within subsystems. A social system emerges wherever two or more actions are connected. At the most basic 'level' Luhmann classifies this as 'interaction'. But as the complexity of these interactions increase, they for-malise into distinct subsystems such as organisations or corporations, each carrying unique specialisation and identity. These societal subsystems self-define 'meaning' and in doing so isolate themselves, creating a unique identity through the selection or rejection of relevant or irrelevant 'communications'.[35]

---

32 Chaos theory has already told us that complete symbiosis is very unlikely.
33 Autopoiesis is a compound word: *auto* meaning oneself and by itself, and *poiesis* meaning production, creation and formation. Hence, the word autopoiesis literally is 'self-production or self-creation'.
34 Varela, F, Maturana, H and Uribe, R, 'Autopoiesis : The Organization of Living Systems, Its Characterization and a Model', 1974, 5 *Biosystems* 187.
35 Luhmann, N, *Soziale Systeme*, 1984, Frankfurt: Suhrkamp.

This process allows an organisation to assume its own 'life', motivated and justified by its selective communication process. In this way, social systems reduce the overwhelming world complexity, establishing difference between themselves (the subsystem) and the environment (all other subsystems).[36] Thus communication is at the heart of Luhmann's theory; subsystems evolve and develop through the internalisation of information communicated from other subsystems. It is my belief that by treating the regulatory matrix as an autopoietic environment, with each group of actors considered a subsystem, we can begin to understand the regulatory environment more fully. In doing so, though, we ask regulators and regulatory theorists to embrace a much more complex regulatory environment, as within Luhmann's model, the effect of each communication between actors is dependent upon the internal logic of each of the external, self-referring subsystems. Control is the fundamental premise of regulation, but within an autopoietic model, control becomes a problem of communication where those subsystems required to implement control are cognitively open, but operatively closed.[37] This means that communications between actors can never be certain, but within Luhmann's terms a communication is a very specific event, allowing us to account for these difficulties in our regulatory model.

In an autopoietic context communication is an 'event' comprised of three key aspects: 'information', 'utterance' and 'understanding', which enable the autopoietic process by way of further communications. Indeed, such communication forms the core of self-referential autopoietic systems and subsystems. Each of these aspects is selected (not necessarily by a person) from numerous possible choices, thereby defining the identity and boundary of the subsystem. Information, as it implies, is the *what* of the message. Utterance is the *how*, the *who* and the *when*. Understanding is the *sense* or *meaning* generated in the receiver. The process of this communication leads to further communications relating to the information imparted, both within the subsystem and potentially within the environment (other subsystems). Through self-reference and the memory of previous selections, a subsystem focuses on only specific communications, as among the possible social connections there are only a few that are relevant or compatible with its identity. Functionally differentiated subsystems within the social systems are thereby concerned and can only be concerned with communications that are relevant to their functioning, autonomous of one another. Thereby communicative acts effectively say nothing about the world that is not classified by the communication itself. This process ensures the creation of highly defined differences and attaches the rationale that identity is the creation of further,

---

36 Luhmann, N, *The Differentiation of Society*, 1982, New York: Columbia UP.
37 Dunshire, A, 'Tipping The Balance: Autopoiesis and Governance', 1996, 28 *Administration and Society* 299.

expected, communications, which form and stabilise boundaries. An entity builds up a unique backlog of selections made and selections negated. It uses this accumulation of selections, its meanings, as values for making future selections. This is a self-referential, closed process that maintains a circular dynamic. Its repetition, over time, maintains the identity and existence of the individual subsystem. As Mingers states:

> We can visualize the whole subsystem as an ongoing network of interacting and self-referring communications of different types and see how they can be separated from the particular people involved. The people will come and go, and their individual subjective motivations will disappear, but the communicative dynamic will remain.[38]

Thus communication in autopoietic systems is not a process directed by the actions of individuals, but is rather a system in which they act as the nodes temporarily located within the communication. People are unable to alter the course of communications as they have formed a self-referential loop within which actors play their part rather than write it. In this way, social systems effectively have a life of their own, which gives direction to the thought and activity of individuals – a communications dynamic helpfully mapped by Mingers in Figure 8.5.

The difficulty with this model is that it only goes part of the way towards solving the problem of designing symbiotic regulatory interventions. It suggests that there are stable patterns of communication within the regulatory matrix, allowing regulators to map the communications dynamic within the

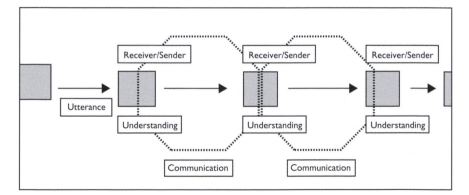

Figure 8.5 Communications as an ongoing process (source: Mingers, *Self-Producing Systems* at p 144).

38 Mingers, J, *Self-Producing Systems: Implications and Applications of Autopoiesis*, *Contemporary Systems Thinking*, 1995, New York: Plenum, p 144.

matrix. This, in turn, allows regulators to anticipate where (and perhaps even when) communication between nodes will take place, suggesting that where known variables can be mapped some nodal responses to the regulatory intervention may be anticipated.[39] Despite this, regulators cannot accurately predict all nodal responses. This is because, as discussed above, the content of communications between actors can never be certain – only the pattern. To actively map the effect of their intervention within the regulatory matrix, regulators must take a further step: that is to measure the probable (or actual) outcome of their intervention through the application of system dynamics.

System dynamics was developed by Professor Jay Forrester of the MIT Sloan School of Management in 1958[40] and is the study of information dynamics within a system, in particular the flow of feedback (information that is transmitted and returned), which occurs throughout the system and the behaviour of the system as a result of those flows.[41] System dynamics starts by defining the problem to be solved. In our example this may be the illicit copying and distribution of copyright protected music or video files. The first step is to information-gather. This requires the regulator to record the current information being communicated by each of the nodes in the matrix, keeping a record of exactly what is being communicated and how. This information, which in our model would have been gathered at stage one – the creation of the autopeotic map of naturally occurring communications – provides a foundational (or first order) model of the system. Using this model as their template the regulator designs a regulatory intervention that they hope will prove to be complementary to the existing regulatory communications within the matrix, leading to symbiotic regulation. The problem is, though, that as the system is complex, it is equally as likely that the intervention will lead to an unexpected response occurring causing one, or more, node(s) communicating either an understanding or information transmission, which could not have been foreseen. The result of such an

---

39 For example, if we return to our example of the Gardener's Dilemma, it means that the regulator can create links or associations between certain actions: knowing that watering the Azalea for instance will have a detrimental effect on the African Violet if it is placed next to the Azalea. Unfortunately he will not know why this is so. To help understand this he must measure the different responses that occur during each change to see which variables cause the change. Although measuring the effect of each change on every component (or node) is computationally intractable, observing the overall effect of each intervention is possible: this is the foundation of systems dynamics.

40 See Forrester, J, 'Industrial Dynamics – A Major Breakthrough for Decision Makers', 36(4) *Harvard Business Review*, 1958, 37; Forrester, J, *Industrial Dynamics*, 1961, Waltham, MA: Pegasus Communications; Forrester, J, 'Market Growth as Influenced by Capital Investment', 9 *Industrial Management Review*, 1968, 105 (among many).

41 For example, system dynamicists study reinforcing processes – feedback flows that generate exponential growth or collapse and balancing processes – feedback flows that help a system maintain stability.

occurrence will be for the intervention to become disruptive. But, by measuring this event, known as feedback, systems dynamics allows for a new, more detailed, second-order model of the regulatory environment to be developed. Thus feedback is both the key to system dynamics and the final piece of our regulatory jigsaw. Forrester explains that decisions, like the environment, are dynamic rather than static. Whereas most decision-makers, including regulators, imagine what he terms an 'open-loop' decision-making process (as seen in Figure 8.6), in truth, decision-making is part of the same self-referential loop outlined by Luhmann and Mingers, meaning that a true depiction of the decision-making process looks more like Figure 8.7.

Information ⟶ Action ⟶ Result

Figure 8.6 Forrester's 'Open-loop'.

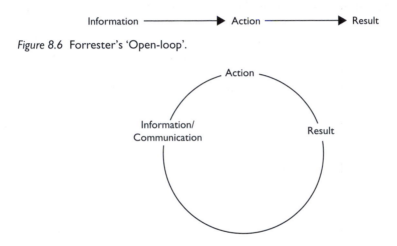

Figure 8.7 Forrester's 'Closed-Loop'.

The key of this 'closed-loop' model is the constant feedback the decision-maker is receiving. Whenever a regulatory intervention is made in any complex environment, whether it be in cyberspace or in a complex real-world regulatory environment, the intervention is scrutinised by all parties and their verdict is communicated to all other regulatory nodes, including the originator of the intervention. This allows the regulator to constantly evaluate and refine their intervention through a process of continual modelling (as seen in Figure 8.8).

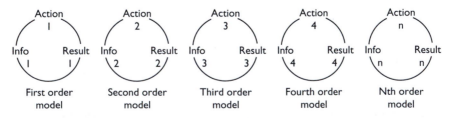

Figure 8.8 Dynamic modelling using system feedback.

At each stage, subsequent to the first-order model, which is designed using the autopoietic map, the regulator is continually amending their actions, based upon the feedback received following their previous actions. Thus Action 1 causes a set of results and resultant feedback, for example adding DRMs to digital media files causes consumer disquiet and a rise in the activity of crackers. As a result the regulator considers this and makes a second intervention – Action 2. This may be attempts to legally control the activity of crackers through legislation such as the DMCA or the Directive on Copyright and Related rights in the Information Society. The effect of this may be to cause a shift in focus from cracking to sharing through file-sharing technologies, leading to a third-order intervention in file-sharing communities and so on. What this demonstrates is that an intervention should not be viewed as a single act, which is then assumed to succeed or fail, depending upon whether it meets a series of subjective standards set by the decision-maker. It, like the regulatory environment, should be dynamically modelled over a period of months, or even years, with each new intervention being designed specifically with regard to the feedback received at each point of intervention. Although this sounds complex, and indeed seems not to be a great advancement on the current model, there are modelling tools such as iThink[42] and Venisim[43] which allow for computer modelling of millions of variables within a digital model.[44] These systems mean that regulators do not need to continue to develop static 'trial and error' regulatory models. They may instead model millions of regulatory variables *before* they make an intervention, suggesting that symbiotic regulation is not something that has to be left to chance or to organic development; by mapping the communications variables within the system and modelling potential feedback patterns using system dynamics it should be possible to achieve regulatory symbiosis on a regular basis.[45]

Regulating cyberspace

Finally, we have a model that goes some way towards describing the complexity of the cyber-regulatory environment, but which also describes how the structure of the environment may be harnessed to provide a more robust

---

42 Developed and supplied by isee Systems. See *www.hps-inc.com/Softwares/Business/ithinkSoftware.aspx.*

43 Developed and supplied by Ventana Systems. See *www.vensim.com/.*

44 The author would like to point out the elegance of harnessing the power of computers to aid in the design of regulatory tools within the complex environment of cyberspace, thus demonstrating that much like regulation, digital technology can be both disruptive and positive.

45 It should be recorded that some US regulators including the Environmental Protection Agency and the Department of Energy now use system dynamics on a regular basis.

regulatory model. At its heart is communication, a discovery that seems rather apt, given that the internet is, after all, a communications device. At the start of this journey I discussed the opposing views of cyberpaternalists and cyber-libertarians. Both believed in their vision of cyberspace and both modelled simple models designed to support their vision. Most famous among these models are David Johnson and David Post's model of an environment without borders and free from state control, and Lawrence Lessig's 'modalities of regulation' model, which described four simple regulatory modalities. Although diametrically opposite in their conclusions, both used a similar methodology. Both assumed a static regulatory universe into which an intervention would be made causing a shift to another static settlement. Regulators would examine this outcome and would declare themselves satisfied or dissatisfied (regulatory failure) and the whole process would begin over. In truth the process of regulation is much more complex. All parties in a regulatory environment continually and simultaneously act as regulator and regulatee. Changes within the regulatory environment are therefore constant and as a result the first stage in designing a regulatory intervention in any complex regulatory environment, including cyberspace, is to develop a dynamic model of the environment, recording all parties and mapping their contemporary regulatory settlements with each other.[46] Second, by observing this environment, regulators are required to map the communications dynamic in place within this regulatory matrix. According to Mingers, the regulator does not need to actually record the content of all communications that take place between subsystems, or nodes; all that is required is that the dynamic of such communication is mapped. In other words the regulator need not anticipate the needs of all actors in the regulatory matrix; they need only anticipate the regulatory tensions that are likely to arise when actors communicate. Finally, once a regulatory intervention has been designed, it should be tested thoroughly. This involves constant monitoring of feedback from all regulatory nodes – both positive and negative. The regulator should be prepared in light of this feedback to make alterations in their position and to continue to monitor feedback on each change, thus allowing them to both accomplish the regulatory settlement they set out to achieve and to generate valuable data, which they may use to model future regulatory interventions.

Effective, symbiotic, regulatory interventions may therefore be designed through the application of a three-stage process. First, regulators must produce a dynamic model of the regulatory matrix surrounding the action they wish to regulate (including a map of the communications networks already in place). From this they may design a regulatory intervention intended to harnesses the natural communications flow by offering to the subsystems, or

---

46 By this I mean which relationships – market, power, social or design cause particular outcomes to occur or not to occur.

nodes, within the matrix, a positive communication that encourages them to support the regulatory intervention. Finally they must monitor the feedback that follows this intervention. If the intervention is initially unsuccessful they should consider modifying it slightly and continuing to monitor the feedback in the hope of producing constant improvements. If successful, the positive feedback generated will reinforce the regulatory intervention, making it much more likely to succeed. If regulators were to use this three-stage design mechanism, it would be possible to design successful regulatory interventions in the most complex regulatory environment. The problem is that a dynamic modelling technique such as this is much more complex to apply during the design phase, and requires constant updating to reflect changes in the dynamics of the communications flow caused by social, economic or technological changes. In addition, this model requires regulators to embrace uncertainty within the regulatory matrix (they may get the outcome they wish but not understand why), and as such requires a remarkable leap of faith. But are regulators, and regulatory theorists, ready to make such a leap?

Chapter 9

Embracing uncertainty

The more precisely the position is determined, the less precisely the momentum is known in this instant, and vice versa.

Werner Heisenberg

In 1903 the physicist Albert Michelson made an astonishing pronouncement. Confident of the advances made by the great physicists of the Victorian era, such as Lord Kelvin, James Clark Maxwell and Thomas Edison, he declared that 'the most important fundamental laws and facts of physical science have all been discovered, and these are now so firmly established that the possibility of their ever being supplemented in consequence of new discoveries is exceedingly remote'.[1] In so doing he followed what appeared to be a long-established tradition that eminent physicists should make bold and far-reaching statements on the nature of the universe and its physical laws.[2] Unfortunately, Michelson's timing could not be worse as early twentieth-century physicists were about to make a string of discoveries that would fundamentally alter our perceptions of the universe. The world as Michelson understood it was about to be turned on its head by the work of three men: Max Planck, Albert Einstein and Ernest Rutherford. The first, Planck, was a respected German physicist who had done work on entropy, but had done little to distinguish himself. Then in 1900 he developed his 'quantum theory' which posited that energy did not flow as a stream or wave as had been imagined before, but rather was carried by packets or quanta. This seemingly simple, yet deceptively complex, thesis was to have a profound effect upon our understanding of the physical universe. Planck's

---

1   Quoted in Coveney, P and Highfield, R, *The Arrow of Time*, 1991, London: Flamingo, p 67.
2   For example only a few years earlier, Lord Kelvin, President of the Royal Society, had stated authoritatively that 'Heavier than air flying machines are impossible'. While Sir William Preece, Chief Engineer of the Post Office and Fellow of the Royal Society described Thomas Edison's incandescent electric lamp as 'an absolute *ignis fatuus*' in a lecture at the Royal United Service Institution in 1879.

thesis proved to be the foundation for the modern discipline of quantum physics, a discipline that first came to the attention of the wider public when in 1905 a young Swiss patents clerk called Albert Einstein wrote five papers for the *Annalen der Physik*. Three of these papers are widely accepted to be among the best in the history of physics and examined photoelectric effects by applying Planck's new quantum theory, Brownian Motion and a 'Special Theory of Relativity'. As Bill Bryson notes in his *Short History of Nearly Everything*, 'the first won its author a Nobel Prize and explained the nature of light (and also helped make television possible among other things). The Second provided proof that atoms do indeed exist – a fact that had surprisingly, been in some dispute. The third merely changed the world'.[3] Thus within two years of Michelson's pronouncement much of our understanding of the physical universe had been turned on its head by this new development; but for Michelson worse was to follow. In 1895, a young South African named Ernest Rutherford had won a scholarship to the world famous Cavendish Laboratory at the University of Cambridge. By 1911, Rutherford had become obsessed with the nature of atoms, the basic building blocks that form the constituent parts of the universe. That year he proposed his 'theory of atomic nuclei', which suggested that all the positive charge and most of the mass of an atom must be contained in a tiny nucleus at the atom's centre, with the negatively charged electrons 'in orbit' about this nucleus. To prove his thesis Rutherford had to 'split' the atom to prove that he could release matter from this nucleus, something which he finally succeeded in doing in 1919 when he showed that by firing alpha-particles into nitrogen gas a small amount of hydrogen could be produced. In so doing he also established that energy could be released from the nucleus of an atom, thereby proving an essential part of Albert Einstein's Special Theory of Relativity: the unification of energy and mass, usually represented by the famous equation $E=mc^2$.[4] With Rutherford providing the link between quantum theory and nuclear physics the Newtonian view of the universe, which was at the heart of Michelson's beliefs, had been rewritten in less that 20 years.

The story of Albert Michelson is a cautionary one: one should never make definitive statements without strong, indeed overwhelming, evidence in support of your position. This though is not why I have chosen to introduce Michelson at this late stage in the text. In fact, Michelson is not even important to the point I wish to make; he merely, better than anyone, serves to illustrate the strength of belief held by eminent physicists throughout

3   Bryson, B, *A Short History of Nearly Everything*, 2003, London: Black Swan, p 159.
4   Einstein, A, 'Ist die Trägheit eines Körpers von seinem Energieinhalt abhängig?' (Does the Inertia of a Body Depend Upon Its Energy Content?), *Annalen der Physik*, 27 September 1905.

the long history of their subject prior to the development of quantum and nuclear physics. Many, including the French physicist and mathematician Pierre-Simon de Laplace, who believed that the universe was completely deterministic and suggested there should be a set of scientific laws that govern everything from the motions of the planets to the nature of human behaviour,[5] believed that Sir Isaac Newton's laws of universal gravitation and motion could provide an explanation for all phenomena. The development of quantum theory displaced this notion, but also led to a much more sophisticated view of the universe, one that suggests the universe is so complex that some aspects of it may be beyond our understanding. At the heart of this complexity lies the infamous 'Heisenberg Uncertainty Principle', developed in 1926 by the German physicist Werner Heisenberg. Heisenberg was attempting to formulate a solution to one of the early problems of quantum physics. The French scientist Prince Louise-Victor de Brogle had discovered that certain anomalies in the behaviour of electrons disappeared if you thought of them as waves, an idea that was then refined and developed into a system of wave mechanics by the Austrian scientist, Erwin Schrödinger (he of the famous cat). At the same time Heisenberg had developed a competing theory called matrix mechanics which seemed to solve a different set of problems to Schrödinger's wave model: they couldn't both be right, or could they? Heisenberg's 'Uncertainty Principle' was a compromise that meant both were right. It states that an electron is a particle, but a particle that can be described in terms of waves. The uncertainty exists because we may know the path the electron takes through space or we may know where it is at any given time, but we cannot know both. Attempts to measure one will disturb the other meaning that both cannot be discovered.[6] The Heisenberg Uncertainty Principle means that quantum physicists must accept that there are areas they cannot measure, or as Professor James Trefil has observed, 'scientists had encountered an area of the Universe that our brains just aren't wired to understand'.[7] Accepting this limitation, twentieth-century physicists set about examining the quantum universe and immediately started to encounter a whole series of events that Michelson and his contemporaries and predecessors could never have imagined. New terms such as 'Quantum Improbability' and 'Quantum Weirdness' were developed to explain

---

5   Hawking, S, *A Brief History of Time*, 1988, London: Bantam. In his most famous quotation Laplace stated, 'An intelligence acquainted with all the forces of nature, and with the positions at any given instant of all the parts thereof, would include in one and the same formula the movements of the largest bodies and those of the lightest atoms'. In reply Stephen Hawking notes, 'We now know that Laplace's hopes of determinism cannot be realized, at least in the terms he had in mind. The uncertainty principle of quantum mechanics implies that certain pairs of quantities, such as the positions and velocity of a particle, cannot both be predicted with complete accuracy.', p.207.
6   This section draws heavily on Bryson, fn 3, p 188.   7   Quoted in Bryson, fn 3, p 189.

phenomena previously thought impossible: phenomena such as 'Quantum Entanglement', which states that certain pairs of subatomic particles when separated by huge distances can instantly 'know' what the other is doing. Thus if you spin one particle in a pair the other immediately starts spinning at the same speed in the opposite direction. This instantaneous communication violates a key principle of Einstein's Special Theory of Relativity: that nothing can travel faster than the speed of light. The problem quantum theories such as this caused led some physicists, most notably Albert Einstein, to regard quantum theory with contempt.[8] Despite such assaults, the study of quantum physics flourished and quantum theory now provides the foundation of our modern appreciation of the universe. Physicists have today embraced the complexity of the universe in an attempt to better understand it. Quantum theory is being applied in the development of everything from computing to cryptography and forms the basis of contemporary attempts to develop a Laplacian 'Grand Unified Theory' such as Stephen Hawking's scientific bestseller *A Brief History of Time* and Brian Greene's *Elegant Universe*.[9]

Physicists had an important decision to make during the first half of the twentieth century. Did they embrace quantum theory with its associated uncertainty, or did they seek to marginalise it? By embracing quantum theory physicists now have probably the most complete understanding of the nature of the universe we have ever had, but at the same time have more questions to be answered than we have ever had; such is the nature of intellectual inquiry. Regulators find themselves at a similar crossroad. Regulatory theorists have begun to develop more sophisticated models to explain the effects of regulatory interventions. One such theory that may soon be seen as regulation's own quantum theory is the theory of decentred or nodal regulation. It is this theory that is at the heart of the dynamic regulatory matrix used throughout this book. It suggests that all actors in a regulatory environment are active participants in that environment: the days of the passive regulatee may be numbered. The significance of decentred regulation is that much like quantum theory, it asks those who practice the discipline to accept uncertainty within their chosen discipline and, further, to embrace that uncertainty as the path to greater understanding. Fulfilling the roles originally assumed by Planck, Schrödinger and Heisenberg are a variety of social theorists

---

8 Famously Einstein, Podolsky and Rosen developed the EPR theory to show the macro-effects of quantum theory, leading to Einstein calling Quantum Entanglement 'spooky action at a distance'. Einstein also famously stated 'God does not play dice' in response to the apparent unpredictability of quantum physics.

9 Greene, B, *Elegant Universe: Superstrings, Hidden Dimensions and the Quest for the Ultimate Theory*, 2000, New York: Random House.

including Niklas Luhmann,[10] Gunter Teubner[11] and Michael Foucault.[12] Their work in autopoiesis and decentred decision-making has formed the basis for detailed study of the effects of multi-nodal regulatory decision-making in several regulatory subject areas. This has been developed by regulatory theorists such as Julia Black,[13] Clifford Shearing[14] and John Braithwaite and Peter Drahos[15] who have suggested that in regulatory subjects as diverse as policing and financial services regulation the effects of decentralisation are clearly observable. This allows us, in the words of Julia Black, to observe the 'complexity of interactions and interdependencies between social actors, and between social actors and government in the process of regulation',[16] a complexity which, much like the quantum universe, has always existed, but which is not measured when one takes a traditional command and control approach to measuring regulation. Thus clear parallels exist between the development of quantum theory, and the development of new regulatory theories such as nodal regulation, dynamic regulatory modelling, communications theory and system dynamics. Regulators, and regulatory theorists, much like early twentieth-century physicists now stand at a crossroads and decisions they make now will influence how we understand of the regulatory universe in the future.

In this book I have demonstrated how such models can be applied in the cyber-regulatory field and further suggest how the application of three simple models can be used to design effective regulatory interventions in this complex environment. The model is similar to models developed from within the John F Kennedy School of Government at Harvard University, in particular

10 Discussed above and at length, Chapter 8.
11 Teubner, G, *Law as an Autopoietic System*, 1993, Oxford: Blackwell; Teubner, G, *Netzwerk als Vertragsverbund*, 2004, Münster: Nomos Verlagsges.
12 Foucault, M, *The Essential Works: Power v.3*, 2001, Harmondsworth: Penguin; Foucault, M, 'Truth and Power', in *The Foucault Reader: An Introduction to Foucault's Thought*, Rabinow, P (ed), 1984, Harmondsworth: Penguin.
13 Black, J, 'Decentring Regulation: Understanding the Role of Regulation and Self Regulation in a "Post-Regulatory" World', 2001, 54 CLP 103; Black, J, 'Critical Reflections on Regulation' 27 *Australian Journal of Legal Philosophy*, 2002, 1; Black, J, 'Enrolling Actors in Regulatory Systems: Examples from UK Financial Services Regulation', [2003] PL 63.
14 Shearing, C and Wood, J, 'Nodal Governance, Democracy, and the New "Denizens" ' 2003, 30 JLS 400; Shearing, C and Stenning, P, 'From the Panopticon to Disneyworld: The Development of Discipline' in Doob, A and Greenspan, E (eds), *Perspectives in Criminal Law*, 1984, Aurora: Canada Law Book; Shearing, C, Wood, J and Font, E, 'Nodal Governance and Restorative Justice', forthcoming, *Journal of Legal Studies*, available at: *www.mj.gov.br/reforma/eventos/conf_internacional/Nodal%20Governance%20and%20Restorative%20Justice.pdf*.
15 Braithwaite, J and Drahos, P, *Global Business Regulation*, 2000, Cambridge: CUP; Braithwaite, J and Drahos, P, 'Ratcheting Up and Driving Down Global Regulatory Standards,' 1999, 42 *Development* 109.
16 Black, 'Decentring Regulation', fn 13, p 107.

Viktor Mayer-Schönberger and John Crowley's self-governance model for virtual worlds,[17] but with the added value that it not only recognises that the complexity and decentralisation of regulatory matrices in the cyber-environment, it creates also the model to allow regulatory theorists to embrace this complexity in a manner not dissimilar to the movement made by physicists in the early twentieth century.

The three-dimensional regulatory matrix, introduced at page 54 and discussed at length in Chapter 8 allows us to imagine the structure of nodal or decentralised regulation in the complex, multilayered, new media environment. By mapping both the horizontal effects of near-perfect communications and the vertical effects of environmental layering in contemporary communications media, we can create a visual representation of the regulatory environment, enabling the regulator to visualise the potential polycentric effects of any regulatory intervention. When this model is then paired with Mingers' representation of autopoietic communication, illustrated on page 246, it allows regulators to chart the communications flow within the regulatory matrix, enabling the construction of a 'first order model' of communication within the regulatory matrix. This first-order model then forms the foundation for an initial intervention and a series of dynamic feedback loops, where the results of every intervention are measured against the responses from each regulatory node, allowing for fine-tuning of the intervention and creating ultimately robust, symbiotic regulation – regulation designed to harness the intrinsic structure of the regulatory environment rather than the disruptive blunt instrument of traditional command and control regulation. These models present a tantalising opportunity for regulators and regulatory theorists. They offer the most complete model of regulatory design and comprehension, but to accept them means to accept the limits of our knowledge as they, like quantum theory, suggest things we cannot know. It is to be suggested that if we are to further our understanding of the regulatory environment within cyberspace (or any complex environment), we must, much like the quantum physicists of the early twentieth century, accept these limitations and use them to our advantage. For knowing what you do not know is as important as knowing what you do.

---

17 Mayer-Schönberger, V and Crowley, J, 'Napster's Second Life? – The Regulatory Challenges of Virtual Worlds', *KSG Working Paper No. RWP05–052* (September 2005), available at SSRN: *http://ssrn.com/abstract=822385*.

Index

Page numbers with n refer to material appearing only in the footnotes.